# INSIDE THE CIVANO PROJECT

## A CASE STUDY OF LARGE-SCALE SUSTAINABLE NEIGHBORHOOD DEVELOPMENT

## McGRAW-HILL'S GREENSOURCE SERIES

**Gevorkian**
*Solar Power in Building Design: The Engineer's Complete Design Resource*

***GreenSource: The Magazine of Sustainable Design***
*Emerald Architecture: Case Studies in Green Building*

**Haselbach**
*The Engineering Guide to LEED—New Construction: Sustainable Construction for Engineers*

**Luckett**
*Green Roof Construction and Maintenance*

**Melaver and Mueller (eds.)**
*The Green Building Bottom Line: The Real Cost of Sustainable Building*

**Nichols and Laros**
*Inside the Civano Project: A Case Study of Large-Scale Sustainable Neighborhood Development*

**Yudelson**
*Green Building Through Integrated Design*

### About *GreenSource*

A mainstay in the green building market since 2006, *GreenSource* magazine and GreenSourceMag.com are produced by the editors of McGraw-Hill Construction, in partnership with editors at BuildingGreen, Inc., with support from the United States Green Building Council. *GreenSource* has received numerous awards, including American Business Media's 2008 Neal Award for Best Website and 2007 Neal Award for Best Start-up Publication, and FOLIO magazine's 2007 Ozzie Awards for "Best Design, New Magazine" and "Best Overall Design." Recognized for responding to the needs and demands of the profession, *GreenSource* is a leader in covering noteworthy trends in sustainable design and best practice case studies. Its award-winning content will continue to benefit key specifiers and buyers in the green design and construction industry through the books in the *GreenSource* Series.

### About McGraw-Hill Construction

McGraw-Hill Construction, part of The McGraw-Hill Companies (NYSE: MHP), connects people, projects, and products across the design and construction industry. Backed by the power of Dodge, Sweets, *Engineering News-Record* (*ENR*), *Architectural Record*, *GreenSource*, *Constructor*, and regional publications, the company provides information, intelligence, tools, applications, and resources to help customers grow their businesses. McGraw-Hill Construction serves more than 1,000,000 customers within the $4.6 trillion global construction community. For more information, visit www.construction.com.

# INSIDE THE CIVANO PROJECT

## A CASE STUDY OF LARGE-SCALE SUSTAINABLE NEIGHBORHOOD DEVELOPMENT

**C. ALAN NICHOLS, P.E., LEED AP**

**JASON A. LAROS, LEED AP**

New York   Chicago   San Francisco   Lisbon   London   Madrid
Mexico City   Milan   New Delhi   San Juan   Seoul
Singapore   Sydney   Toronto

**The McGraw-Hill Companies**

Cataloging-in-Publication Data is on file with the Library of Congress

Copyright © 2010 by The McGraw-Hill Companies, Inc. All rights reserved. Printed in the United States of America. Except as permitted under the United States Copyright Act of 1976, no part of this publication may be reproduced or distributed in any form or by any means, or stored in a data base or retrieval system, without the prior written permission of the publisher.

1 2 3 4 5 6 7 8 9 0   DOC/DOC   0 1 5 4 3 2 1 0 9

ISBN 978-0-07-159931-3
MHID 0-07-159931-2

**Sponsoring Editor:** Joy Bramble Oehlkers
**Editing Supervisor:** Stephen M. Smith
**Production Supervisor:** Richard C. Ruzycka
**Acquisitions Coordinator:** Michael Mulcahy
**Project Manager:** Virginia Howe, Lone Wolf Enterprises, Ltd.
**Copy Editor:** Jacquie Wallace, Lone Wolf Enterprises, Ltd.
**Proofreader:** John Snyder
**Art Director, Cover:** Jeff Weeks
**Composition:** Lone Wolf Enterprises, Ltd.

Printed and bound by RR Donnelley.

McGraw-Hill books are available at special quantity discounts to use as premiums and sales promotions, or for use in corporate training programs. To contact a representative, please e-mail us at bulksales@mcgraw-hill.com.

 The pages within this book were printed on acid-free paper containing 100% postconsumer fiber.

---

Information contained in this work has been obtained by The McGraw-Hill Companies, Inc. ("McGraw-Hill") from sources believed to be reliable. However, neither McGraw-Hill nor its authors guarantee the accuracy or completeness of any information published herein, and neither McGraw-Hill nor its authors shall be responsible for any errors, omissions, or damages arising out of use of this information. This work is published with the understanding that McGraw-Hill and its authors are supplying information but are not attempting to render engineering or other professional services. If such services are required, the assistance of an appropriate professional should be sought.

# Dedication

In memory of Carol Goodwin, Civano's first resident.

December of 1995 was a critical time for Civano. The outcome of a single City Council vote was to seal the fate of the Solar Village. Carol Goodwin, a strong supporter of the Solar Village concept, stood at the podium and insisted that she would be the very first to reside in the Village. With her support, and the support of other people with such conviction, the Solar Village dream survived. Sadly, Carol did not survive to see her home built, but her mortal ashes were freed in Civano at the lot she had purchased. Today, Carol is remembered by a tree in the center of the Neighborhood Center courtyard.

# About the Authors

**C. Alan Nichols, P.E., LEED AP,** established Al Nichols Engineering in 1995. He has served as Project Engineer at Western Electric, Process Engineer for W. L. Gore, and Project Engineer for Tierney Manufacturing. Nichols has over 30 years' experience in heating, air conditioning energy systems, and plumbing. As a member of the Tucson/Pima County Metropolitan Energy Commission, he was instrumental in writing the sustainable energy standard (SES) for Civano. Additionally, Mr. Nichols was part of a volunteer group that led the development of building code guidelines that have resulted in Civano's 50% reduction of heating and cooling energy and 60% reduction in potable water usage. In 2002, he received the Energy Users News Award for Best Mixed-Use Facility for the Civano project, and he is Past Chairman of the Tucson/Pima County Metropolitan Energy Commission.

**Jason A. Laros, LEED AP,** is Sustainability Analyst at Al Nichols Engineering as well as Membership Committee Chairperson and Governing Council Member for the United States Green Building Council, Southern Arizona Branch.

# CONTENTS

**Foreword**     *xi*

**Preface**     *xiii*

**Acknowledgments**     *xvii*

**Chapter 1    A Seed Is Planted: The Early Stages of Land Planning**     **1**

    Planning Permits    *4*
    Conforming    *9*
  INTRODUCTION TO THE CHARTER OF THE NEW URBANISM    *11*
    Location, Location, Location    *12*
    Proximity to Water and Wastewater Infrastructure    *13*
  FRAMING THE COMMONS    *14*
    Imperiled Species and Ecological Communities    *14*
    Wetland and Water Body Conservation    *15*
    Useful Agricultural Land Conservation    *15*
    Floodplain Avoidance    *16*
    Brownfield Redevelopment    *16*
    High Priority Brownfield Redevelopment    *17*
    Preferred Locations    *17*
    Reduced Oil-Fueled Automobile Dependence    *17*
    Bicycle Infrastructure    *18*
    Fantasy Island    *18*
    Local (Walking/Bicycle Scale) Housing and Jobs    *20*
    School Proximity    *20*
    Solar Radiation, Wind and Rainfall    *24*
    Back to the Plan    *24*

**Chapter 2    The Public/Private Partnership: A Balancing Act**     **27**

    The State Open Public Meeting Law—Title 38, Chapter 3, Article 3.1    *29*
    Land Development Meetings    *31*
  REFRAMING THE TRAGEDY    *38*
  OIL OVERCHARGE FUNDS    *39*
    Civano Gets a Petrol-Funded Boost    *40*

    The Tucson Institute for Sustainable Communities    44
    Pre-Planning    45
STATE LANDS DEPARTMENT    46
URBAN LANDS ACT    47
METROPOLITAN ENERGY COMMISSION    48
    Ideal Ideas    49

## Chapter 3    Guiding Growth: Master Planning and Analysis    51

EXCERPT OF PAUL ROLLINS' INTERVIEW WITH WAYNE MOODY    53
IAN MCHARG'S *DESIGN WITH NATURE,* 1969    58
MASTER PLANNING: APPLIED NEW URBANISM    61
    Consider Density    61
    Diversify Uses    62
    Mixed Use Zoning—Coming Back Home    64
    Transportation Network    65
    Connect to Surroundings    68
    Public Spaces    68
    Maximize Accessibility for All    69
    Involve the Community    69
    The Plan Book That Never Was    74
    New Urbanism's Code    75

## Chapter 4    Impacts and Adjustments: The Basics of High Performance, a.k.a. Green    77

AUDITING THE PROJECT, ENERGY AND WATER    80
CHARACTERISTICS OF THE 2007 ENERGY AND WATER USE MONITORING STUDY    81
    Correlations    82
    Evaluation of 2007 Energy Use    84
    Cost and Energy Savings for the City of Tucson and Civano    88
WATER USE    89
    Civano Phase I Water Use in Common Areas    90
CIVANO PHASE I VS. PHASE II COMPARISON    91
    MOU Adaptation    91
HOME RESALE VALUES: PHASE I, PHASE II, AND A NEIGHBORING DEVELOPMENT    92
INTER-CIVANO ENERGY COMPARISON    93
INTER-CIVANO WATER COMPARISON    94
LEED—NEIGHBORHOOD DEVELOPMENT "CERTIFICATION OF A COMPLETED NEIGHBORHOOD DEVELOPMENT"    95
    Embodied Energy    97
A CLOSER VISION    99

## Chapter 5    Germination    *101*

    THREE MAJOR LESSONS LEARNED    *105*
        #1 Lesson Learned: Capitalization    *105*
        #2 Lesson Learned: Size Matters (Capitalization Continued)    *108*
        #3 Lesson Learned: Identify Goals and Send a Clear Message    *110*
    PUBLIC MEETS PRIVATE: THE CIVANO INSTITUTE    *113*
    LEE RAYBURN SUMMATION    *120*

## Chapter 6    Tug of War: Rediscovery    *121*

    CHOOSING A MASTER DEVELOPER    *125*
    MARKET STUDY FOR CIVANO    *128*
    WHY FEAR THE "S" WORD?    *131*
        Selling Civano    *135*
        Goodbye Fannie Mae    *139*
    DEFINING AFFORDABLE HOUSING    *144*
        Compounding Effects of Energy Efficient Building Envelopes    *145*

## Chapter 7    Ground Breaking: Neighborhood One    *149*

    URBAN DESIGN FEATURES SPECIFIC TO PHASE I CIVANO    *150*
        The Streetscape and Landscape    *151*
    THE IMPACT PROCESS, CONTINUED    *152*
    BUILDING MATERIALS, TECHNIQUES, AND TECHNOLOGY ADOPTION    *157*
    SOLAR POWER IN CIVANO PHASE I    *157*
        Solar Electric    *158*
        Residential Scale Solar Thermal Technology Overview    *159*
        Solar Water Heating Lessons Learned    *162*
        Wall Systems    *162*
    RECYCLING IS A COMMUNITY PERFORMANCE METRIC    *166*
        Heating, Ventilation and Air Conditioning (HVAC) in Civano    *169*

## Chapter 8    A Middle Ground—Phase II: The Costs and Benefits of Production Housing    *175*

    THE COMMUNITY CENTER    *178*
        Affordable New Urbanism    *188*
        Speed of Buildout    *190*
        Urban Villages    *191*
        Did the Public and Private Partnership Help or Hurt?    *193*
        Does Sierra Morado Meet Project Goals?    *195*

## Chapter 9    Civano's DNA: Leading the Evolution    *199*

    Lessons Learned    *199*
    Architects, Engineers, and New Urbanists    *209*

THE ENGINEER, IN HIS LANGUAGE   *210*
   2005 Sustainable Energy Standard (Appendices A and B)   *214*
   Conclusions for the 2007 Energy Use Study   *215*
LEED—NEIGHBORHOOD DEVELOPMENT, GREEN CONSTRUCTION AND TECHNOLOGY   *216*
   Construction Activity   *216*
   Green Construction   *217*
   Energy and Water Efficiency   *217*
   Minimize Resource Consumption and Site Disturbances   *219*
   Managing Stormwater   *219*
   Solar Management   *219*
ONSITE ENERGY AND RENEWABLE ENERGY GENERATION   *222*
   District Central Plants   *223*
   Wastewater Management   *224*
   Recycled Content in the Infrastructure and Waste Management   *226*
   Dark Skies   *226*

**Chapter 10   The Future Neighborhoods: Phase III, IV, the Commercial Center, and Beyond**   **229**
   Fast Forward   *231*
   The Civano Institute: Revisited   *233*
   Can Future Buildings Be the Environment?   *235*
   The Future   *236*

**Epilogue**   **241**

**Glossary**   **243**

**Appendix A   Civano Impact System Memorandum of Understanding on Implementation and Monitoring Processes; Signed June 26, 1998**   **251**

**Appendix B   Revised Sustainable Energy Standard**   **287**

**Bibliography**   **295**

**Index**   **297**

# FOREWORD

The following story is about a group of people who recognized a once-in-a-lifetime opportunity and took it upon themselves to reinvent the future and work for its realization. The great vision of this project was understood long before those still preoccupied with the status quo could accept its premise. Yet, the story unfolds despite all the resistance and pitfalls in its path. While the final outcome was never assured, the dream and ultimate value of this opportunity would never be shaken. And so this is the story of the Tucson Solar Village Project later to be renamed Civano.

The 1970s was a decade much like the present period–protracted energy crisis, a long global recession, major changes in our monetary system, high unemployment, conflict and wars, heightened global competition for resources and markets, inflation, and growing interest in solar energy and new, more efficient technologies. In short, the world was burdened with uncertainty but also drawn to the possibilities of great and needed change.

The key scientific finding during the seventies was that Planet Earth is a finite system and that unconstrained growth would lead to dangerous resource shortages and degradation of natural systems. When the bad news would arrive in the future—when the limits to growth would be reached—was debated endlessly among "experts." However, in everyday terms, these limits seemed so far off in time that more immediate, pressing issues absorbed most people's attention. Except for a relatively small community of scientists and environmental thinkers, the value of alternatives to growth was very low. Why pursue sustainability when the presumed nature of the economy and the world's capacities is to always grow.

In the United States, the seeds of the sustainable development movement germinated in the seventies in response to these scientific findings and the social and cultural fallout from economic instability. For those of us who came to see this coming paradigmatic change, many would have to carry this knowledge patiently into the eighties and nineties before the time was right to actually work on planning, engineering, and building a different world.

The story of Civano is a story of many diverse people and events coming together at different points in time to move forward the proposition that now is the time for a prototype sustainable community development. The designs changed and evolved, but the vision was always a comprehensive treatment of all functions of the human built environment in harmony with the natural cycles of energy, water, materials, and eco-systems.

When this opportunity appeared to the first wave of Tucson innovators in the 1980s, it was clear that the next evolutionary phase was beginning. We would lead the first major experiment in the desert Southwest for learning how to create a community land development based on regenerative cycles and significantly reduced resource consumption. The promise of wide-scale utilization of solar energy in the future would be furthered by this single venture in a new approach to development.

The chronology of Civano spans three decades with important milestones achieved in each. At many points, the realization of the dream stood in doubt as challenges overwhelmed the participants and the institutions backing its progress. But key actors always kept the effort moving forward up to its current state as a living, breathing place where people and families live their lives.

An experiment should never be labeled either a success or failure because the underlying purpose of an experiment is to test hypotheses and learn about something which often has never been attempted before. Civano provides us with a unique set of valuable lessons for designing ongoing responses to the intensifying sustainability crises unfolding all around us.

The story of Civano is ultimately a story of local heroes carrying forward a noble and important mission. In particular, I want to acknowledge Al Nichols, engineer extraordinaire, for his many roles throughout the past two decades in bringing Civano into being and making its beneficial lessons available to all those now and in the future who will take on the next critical sustainability challenges.

*Robert Cook*
*Former Chair, Tucson/Pima County Metropolitan Energy Commission*
*Co-Founder, Sustainable Tucson*

# PREFACE

*What's in a name? That which we call a rose by any other name would smell as sweet.*

—SHAKESPEARE, *Romeo and Juliet,* II ii 1–2

The quote above may have been true in the century William Shakespeare wrote, but in the current age of information, labeling matters. A community called Military Compound indicates a distinctive purpose and ambiance which differs widely from the expectations for a community named Green Meadows. If we follow the history of names given or associated with the project known as Civano from its inception in 1982 to the present fruition in 2009, we find a richness of meaning in parallel stories and an evolution of the purpose from a singular ideal to a complex of ideology and practicality.

The Tucson Solar Village was conceived as a solar demonstration project on the scale of a community rather than a single residence. The goal for the project as envisioned was straightforward—to build a housing project which showed how to use the abundant solar energy of the Sonoran desert to reduce grid energy use. Over the next decade the purpose for the project grew. Civano began took shape through a unique planning process which included private citizens, businesses and public officials from government and universities. By the time this project was officially named Civano the purpose had grown to encompass a complex blend of sustainable lifestyle ideals and the demonstration of community design for high efficiency green building techniques.

"Civano" is the name of a historical phase of a remarkable Native American civilization, the Hohokam, who inhabited this area 700 years ago. Parts of the Hohokam history parallel our present day challenges and gives deeper context to the namesake project. This Civano Phase was an environmentally difficult time following a comparatively less challenging period of expansion in population and cultural sophistication. Recurrent flooding and drought compounded challenges the Hohokam were experiencing beyond their ability to maintain their organization structure or centralized authority. The Civano Phase marked a decline of the sheer extent of Hohokam development, but it was not a decline of their civilization. Faced with new social and environmental circumstances, the culture adapted. The name Civano was derived from a word which indicated the chief of a great house. Stories indicate that it was such a chief who helped foster a new way of thinking—a surge of innovation and a departure from earlier Hohokam traditions.

Civano Phase I now stands as the pre-eminent example of how to develop an entire growth corridor during a time barraged by environmental and economic challenges. To date, approximately 10,800 acres are under master planning. The requirements for energy efficiency and solar use have spread as Sierra Morado by Pulte Homes, The Orchards by Pepper Viner and The Presidio by Ducette continue to meet the standard for high efficiency building at a community scale established by the first Civano planners.

The Houghton Area Master Plan (HAMP) planning effort is largely modeled after the Civano project—with the significant exception of the environmental standards which, to date, have been omitted from the HAMP requirements. As studies about affordability and life-cycle costs of high performance buildings utilizing solar energy become apparent, there is a desire by the Metropolitan Energy Commission to make the Houghton corridor follow the Civano environmental standards as well.

This is the story of the inception, planning and performance of Civano; a master planned, high performance community. The state of the project is monitored by yearly audits which have yielded data and lessons learned which can be used to help citizens, building officials and developers relate to the burgeoning industry of high performance (a.k.a. green) building. Civano is the grand experiment which demonstrates affordable and increasingly sustainable development practices that incorporate renewable solar energy; for living soundly into the future.

If our planet were a beach, Civano would be a mere pebble, but move a pebble where no one has before and it will perhaps create a cascade of consequences which will alter the universe. While we continue to develop a remarkable global civilization we move mountains in an attempt to mold our environment to meet our needs, wants and desires. An old adage, "Necessity is the mother of invention," reminds us it is logical to move forward as if we have the potential to take a mere pebble, a seed of thought, and nurture it as though these are the early motions which will lead to the natural outcome of our civilization. But with each new tool we wield, our actions increasingly mold the future and our conquests of nature play a real and measurable role in altering the course of earthly environmental evolution.

We believe there have been defining moments in the history of our civilization when a single vision—a Civano—has changed the inevitable outcome with hope for a better future.

As resources continue to become more and more limited, a grand plan such as the Civano project is a rational progression towards sustainability that takes advantage of our most abundant renewable energy resource, solar energy, while reducing oil-fueled transportation and consumption of our irreplaceable water resources. Although the journey to Civano began in 1981, systematic approaches to green building are only now becoming mainstream. There are numerous programs springing up around the country. Some are developed by municipalities like Santa Monica or Scottsdale.

Leading this transformation are professional organizations like the National Association of Home Builders (NAHB) and the United States Green Building Council (USGBC). The USGBC's LEED—Neighborhood Development pilot program is the only program being specifically developed for community-sized building projects at the time of this writing, but it takes the advances being made on all fronts of the building industry to help transform our built environment and nurture a new vision of how we will live more sustainable now and on into the unforeseen future.

Civano demonstrates the marketability of sustainable community development on a large scale at affordable prices. This 820 acre traditional neighborhood development utilizes proven, available technology to reduce natural resource usage substantially below current levels. The property is located on State Trust land in the City of Tucson, southeast of Houghton and Irvington Roads, where special zoning was approved to support the Civano project.

Civano addresses the growing desire for a new development pattern that enables people to meet their economic needs, yet maintain social values and ecological harmony. Civano, with an approximate population of 2,500 at the time of this writing, will become home to more than 5,500 people and the location of light industry, offices, a hospital, schools and retail businesses. Commercial, cultural, and civic activity clustered in the village center will foster a small town ambiance. The New Urbanism zoning goal is for half the population and two-thirds of the jobs to be located within a five minute walk of Civano's downtown commercial center.

Civano's master plan envisions construction of approximately 2,500 homes using significantly fewer natural resources than conventional homes. The challenge to builders is to develop housing with these features in a wide range of prices. This goal recognizes both the public policy intent of the plan and the economics of the competitive market area. The City of Tucson supports innovative approaches to these issues. To attain these goals, aspects of regional development, community design and building science must work in harmony.

Potable water reductions were gained using the technology of the City of Tucson's reclaimed water distribution system and the design of residences with smaller yards and community xeriscape. Reductions in home heating and cooling energy is achieved through a proactive building code called the Sustainable Energy Standard (SES)—which leads to the most practical utilization of available technologies that help achieve that goal (added insulation, high efficiency HVAC equipment, solar water heaters, excellent windows). Zoning was used to make commercial and mixed use areas within Civano to create the one job/two resident potential and place services within the community. This serves to lessen vehicular travel and the impacts associated with the traditional sprawling neighborhood. One of Civano's most outstanding examples is the mandatory use of solar energy to at least meet minimum Civano code requirements. At the early time of this writing a large private developer, WESTCOR, was in negotiation to take over the Houghton Area Master Planned site—as well as additional surrounding acreage. Will they use this opportunity to adopt the Civano model, including the Sustainable Energy Standard, and develop what would undoubtedly be the world's largest high-performance master-planned development project? Civano was a hard-earned stepping stone born of a public/private partnership of citizens, professionals and officials who utilized advanced planning that encompassed issues of ecology, energy, resource scarcity and community. It is from this stance that people can change and influence a very old existing paradigm that promotes sprawling bedroom communities to the detriment of our society.

So, as work on the HAMP continues, will the citizens and officials of Tucson move once again to require these standards are met or exceeded? Why wouldn't they? The developers and builders at Civano took the original risk. By their 25-year efforts, the groundwork has been laid, the foundations have been poured and the homes have been occupied. Indeed, the numbers are coming in, the builders have been making money, and the buildings are running lean. The gauntlet has been thrown down.

Whatever comes to pass today, years from now when we walk along Civano's stretch of South Houghton Road in Tucson, Arizona, we will be able to look at the structure of the buildings there and know exactly how they function the way they do; how their walls, roofs and windows derive resistance to heat transfer and how they use solar energy to heat water, make electricity and light buildings. We will see the various urban design approaches: mixed use zoning, the walking paths, open spaces, beautiful low-water-use desert landscape and narrow streets that make it such a pleasant environment to be in. What we must carry with us as we walk away are Civano's seeds: the original intentions of the men and women who built it—so greatly improved over the developments that came before, and at little or no additional cost.

*Inside the Civano Project* includes the general steps that must be considered before and during any development projects that alter the built and natural environment—much of the language in this 26-year story has been guided by the current programs and trends that are becoming mainstream. This work provides insight into those elements of the planning of Civano that have proven fruitful and those that could be improved upon; for Civano came before its time.

As you read, please remember that green is a color, and sustainability is an ideal. Although we use these terms in this book, the search for sustainability is conducted by making each generation of the economy and the built environment higher performance than the last at a rate that exceeds our demands. As of yet, modern practice and technology are anything but sustainable—only improving a little bit at a time. Such paradigm shifts are a slow process, but like the Hohokam civilization once was, we are faced with many challenges and may be running out of time.

# ACKNOWLEDGMENTS

This book had many contributors, and we did not come close to reaching everyone. Special thanks go to Justin Cupp, Home Improvement and Maintenance (HIM), for his artistry and illustrations; Gina Burton-Hampton, for being our in-house book "architect"; and Brian McDonald, the fearless, zero-carbon intern.

Interviewees (in alphabetical order):

Richard Barna
Pam Bateman
Simmons Buntin
David Butterfield
David Case
Bob Cook
Doug Crockett
David Elwood
Rick Hanson
Karen Hidel
Vinnie Hunt
Kevin Kelly
Katharine Kent, P.E.
John Laswick
Hector Martinez
John Miller
Wayne Moody
Valerie Raluk
Paul Rawlins
Lee Rayburn
Sandy Reichhardt
Shirley Scott
Less Shipley
Jim Singleton
Bill Webber
Ardi Whalen
Gal Whitmer
Martin Yoklic
Jerry Yudelson

# INSIDE THE CIVANO PROJECT

## A CASE STUDY OF LARGE-SCALE SUSTAINABLE NEIGHBORHOOD DEVELOPMENT

# A SEED IS PLANTED

## The Early Stages of Land Planning

*We know more from nature than we can at will communicate...*

—EMERSON

By 1850, only 2 percent of the world's population lived in cities of 100,000 people or more. By 1950, the combined effects of industrialization and extended life spans led to an increase of city dwellers by over 15 percent. By the time ground was broken in Civano just a few years before 2000, approximately 50 percent of all people lived in large cities and two-thirds lived in urban regions (Weeks, 1999). Throughout the twentieth century, city growth expanded at a much faster rate than total population growth. As people increasingly poured into urban centers the population growth put stress on aging infrastructure, creating environmental damage and leaving our urban centers with inadequate systems in constant need of upgrading.

Bridges that were built decades ago, or even a century ago, are bearing new loads—perhaps uncalculated loads. Old, inefficient buildings far outnumber new units. Imagine just the chore of replacing every roof in the nation. How many metric tons of waste does that comprise and how many new materials just to keep roofs over our heads? The power grid is constantly taxed by new growth and the need to replace archaic systems, but we are still grid dependant. Deferred maintenance puts any development routine into a mode of expensive retroactive repair rather than proactive preventative maintenance and value-adding retrofits. It was under these circumstances of the later twentieth century that a small group of visionaries came together in the desert of southern Arizona with the goal of meeting the challenges of housing a growing population through careful planning to build a more sustainable future.

Currently, 23 years after the Land Committee labored to identify a suitable portion of state trust land to develop a green demonstration community, there exists a budding growth of prescriptive and performance-based methods for the planning of increasingly sustainable buildings and development practices. On the forefront of the effort to increase sustainable building practice is the United States Green Building Council (USGBC)—a non-profit organization dedicated to the positive market transformation toward the development of a high-performance (a.k.a. green or sustainable) built environment.

**Figure 1.1**
An artist's depiction of the Hohokam's Path of Life, an ancient petroglyph.
*Drawing courtesy of Justin Cupp.*

The USGBC continually develops a green-building rating system intended to accelerate global adoption of increasingly sustainable building and development practices called the Leadership in Energy and Environmental Design (LEED) Green Building Rating System. The LEED system is intended to encourage and accelerate global adoption of sustainable green building and development practices through the creation and implementation of universally understood and accepted tools and performance criteria.

The majority of green building programs focus mostly on the building structure and site alone. Leading the charge, the USGBC is moving toward a larger-scope view by creating a program for guiding the development of increasingly sustainable community-sized developments. The USGBC's Leadership in Energy and Environmental Design for Neighborhood Development Rating System (LEED—Neighborhood Development), in its pilot period during late 2008, is the first national system for neighborhood design. The LEED—Neighborhood Development program was fruit born of collaboration between the USGBC, the Congress for the New Urbanism, and the Natural Resources Defense Council. Like the Civano process, LEED certification required a more holistic view of the built environment to address issues of location and linkage; neighborhood pattern and design; green technology and construction; and innovation in design process.

The scope of the budding LEED—Neighborhood Development system requires participants to go through a three-stage certification process which provides an optional review of a proposal, certification of an approved plan, and certification of a completed neighborhood development (in accordance with the approved plan). For other LEED programs which focus on new or existing buildings and campus-sized developments, independent, third-party verification of post occupancy performance is required to substantiate that a building or development's design meets accepted high levels of environmentally responsible development.

As of this writing, this process has not been established for the LEED—Neighborhood Development program. However, monitoring is a crucial component of any performance-driven project to determine the extent of successes and the shortcomings that must be addressed. Enter Civano—a public and private partnership that *requires* third party, post occupancy verification of performance in order to substantiate that project goals are being met, and requires adjustment accordingly if they are not.

Although the methodology for developing Civano came before the LEED—Neighborhood Development program, the USGBC's program can provide an interesting, and very pertinent, point of reference from which to analyze Civano. Civano, in turn, has been audited for several consecutive years and is unique as a large scale project yielding real data which may be used to extrapolate some of the benefits to be realized by developers who utilize the LEED—Neighborhood Development program or other programs aiming for similar goals and using like methodologies.

Throughout this book, there are sections that contain a basic analysis of how Civano developed, drawing talking points from LEED—Neighborhood Development and other green building program expectations. This analysis is not intended as a critique of the LEED—Neighborhood Development, NAHB Model Green Home Building Guidelines, or any other green building system. The intention of these comparisons is to use the United State's first industry-scale rating systems and language for creating a high performance built environment and pair it with the story of the United State's largest scale high performance master-planned community to date yielding performance data. The authors hoped to bring to the forefront the basic steps and considerations that must be explored while designing a master-planned, high-performance community. More importantly, the comparison offered a set of "lessons learned" during the development of Civano that may be applied to other projects—at a much more palatable cost.

And so our story started nearly 30 years ago. One day in 1981 Arizona Governor Bruce Babbitt received a private tour of the exhibits on the first Solar Parade of Homes in Tucson. Impressed, Governor Babbitt complimented his guide, John Wesley Miller, and asked him, "What are you going to do for an encore?" Loving a challenge, Miller took the question as a friendly test and immediately started thinking about the answer; but weary from 23 consecutive days of work leading up to the Solar Parade of Homes, John took his thoughts with him on a hard-earned vacation. Just as it may be impossible to know exactly which singular event leads to a particular outcome, John Wesley Miller did not yet know he would later share a vision with Governor Babbitt—and in doing so, plant the seed that has become **Civano** (*a.k.a.* "The Tucson Solar Village").

**John Wesley Miller:** Builder and Solar Advocate

It was a dream I had on a beach in Puerto Vallarta. We had just completed the first Solar Parade of Homes in Tucson, 1981, for which I was the chair. So, I went off on a little vacation, because it had been kind of intense during the Parade of Homes, working every day for three weeks and four weekends. Getting ready for it was quite a challenge for all the builders. On vacation, I was reading a book called 'The Golden Thread' about the history traced back to the Roman days and how solar energy was utilized for baths and other things. When they burned up all the wood they looked for other energy sources and they perfected a lot of really nice solar systems. They simply put black slate on the slope facing the south and ran water over it. The sun heated the black slate and the water went down into the roman baths and the baths were literally heated by solar energy. So, it is not a new science.

As I was reading "A Golden Thread," it described developing land and building homes with solar as a focus. I may have read more into it than was there, but my focus had been on solar seriously since about 1973. So, in 1981 was an opportunity to get the interest of the local builders and do something to demonstrate what we could accomplish with solar in housing. So, out of "The Golden Thread" came the idea to ask the governor to set aside a section of state land so we could really comprehensively plan a development that would include everything, including jobs.

And so it goes, John Wesley Miller decided what his encore might be: to identify a selection of state trust land on which to devise a master planned, high performance neighborhood that would be a demonstration model of what could be achieved using integrated urban planning and building design processes. Miller brought his idea back to the Solar Energy Commission Chairman, Carl Hodges, then-director of the University of Arizona's Environmental Research Laboratory. Together, Hodges and Miller presented their idea to Governor Babbitt who, immediately in support of the idea, introduced Miller and Hodges to the Arizona State Land Commissioner. It was by the recommendation of the Arizona State Land Commissioner that they decided to use the planning permit process which would allow for planning to commence for development of state land prior to the sale of parcels to developers.

## PLANNING PERMITS

The Arizona State Land Department uses the common practice of establishing guidelines for the scope of work under a planning permit. While the final scope of work for individual planning permits is negotiated with the successful applicant, the guidelines attempt to ensure the best interests of the land department and protect the designated trustees. Generally, work under a planning permit undergoes a three-phase process, overseen by the State Land Department (SLD). The guidelines that were imposed in 1985 by Arizona's SLD for the planners of Civano would be similar to those in use currently. The outline of work expected by the successful planning permit applicant for the Solar Village was as follows:

### Phase I – Site and Area Analysis

1  A site analysis to include
   a  Accurate boundary survey
   b  Base map with area calculations
   c  Topographical map and slope analysis
   d  Preliminary soils and geology investigation and analysis (to identify constraints to development)

  **e** Preliminary drainage and hydrology investigation and analysis specifically concerned with identifying flood-prone areas, drainage patterns, and flow requirements

  **f** Existing and planned utilities and infrastructure investigation to anticipate the scope of utility expansion to the proposed development site

  **g** Environmental site analyses-which may include archeological, visual, auditory, vegetation, existing buildings, and climatic factors.

**2** An area analysis relative to the potential development site that considers:

  **a** Surrounding land uses

  **b** Existing general plans (such as the Houghton Area Master Plan)

  **c** Existing zoning and rezoning for the two years proceeding the permit request

  **d** Area development trends

  **e** Existing and planned land use with emphasis on major projects with higher levels of impact to state trust land.

**3** Meetings

  **a** To involve State Land Department (SLD) staff and stakeholders

  **b** To discuss methodologies prior to work and again later to track progress and analyze the products of the planning work.

**4** Prepare a site and area analysis report to document information obtained in the first phase of planning (phase I) and to outline the physical opportunities and constraints of the proposed development site.

**5** Site area analysis report including the following deliverables:

  **a** Location map

  **b** Aerial photo of the site

  **c** Topographical analysis of the site

  **d** Area map

  **e** Survey and drawing of the parcel

  **f** Development constraints and opportunities

  **g** Visual survey

  **h** Vegetation analysis

  **i** Soils analysis

  **j** Surrounding land uses

  **k** Zoning

  **l** Utilities

  **m** Streets

  **n** Usually a number of copies of the report, and a reproducible master will be requested as part of the submittal for phase I.

## Phase II – Market Analysis

**1** Define the limits of the study market area

**2** Analyze the current and future socioeconomic trends of the market area including consideration for at least:
   **a** Population
   **b** Income
   **c** Employment
   **d** Age
   **e** Education
   **f** Length of residence.

**3** Provide an overview of market categories in the context of the metropolitan region in question.

**4** Indentify and analyze markets for major categories of land use for at least:
   **a** Competitive position of study area for major land uses and activities
   **b** Residential
   **c** Office
   **d** Commercial
   **e** Resort
   **f** Hotel
   **g** Other or special use.

**5** Review economic and real estate trends and projections by immediate site area and broader market area for items such as:
   **a** Location of new and proposed developments by land use type
   **b** Building permits by land use type
   **c** Housing conditions and characteristics
   **d** Absorption trends and patterns by land use type
   **e** Occupancy characteristics (multi-family versus single family or strip mall versus shopping center development, etc.)
   **f** Inventories of space and vacancy rates by land use type and occupancy type
   **g** Land values and relative rental rates
   **h** Traffic counts.

**6** Recommend a physically and economically feasible land use mix that gives the highest and best return to the trust. This must include a discussion of the rationale for the amount and type of each proposed land use and include the marketability study findings for each situation.

7. Recommend a phasing schedule for each identified land use.

8. Meetings
   a. Prior to beginning work on phase II to discuss methodologies and objectives of the market study
   b. Follow up after completion of phase II planning to review facts and findings of the market study.

9. Prepare a market analysis report to include the documentation and findings from phase II research and planning. This may outline the suggested land use mix for use in preparing alternative and final plans.

10. Provide copies of the report with a replicable master to the SLD.

### Phase III – Preparation of Alternative Plans and a Final Plan

1. Alternative Plans:
   a. A land use plan report to include three alternative patterns of land use and urban activities including the acreages and densities of each proposed use
   b. Circulation plans (for each alternative plan) that will consider the movement of people and goods throughout the study area-safety and efficiency are paramount concerns
   c. Each alternative plan should include a public facilities, utilities, and services plan to consider the public infrastructure (other than transportation) which is necessary for efficient functioning of a community.

2. Plans:
   a. The three alternative plans need to be based on the analysis of phases I and II to represent the most economically feasible and best uses for the site. Included should be consideration for energy conservation considerations, including but not limited to proposed lot and building orientations, street alignments, and shadow patterns.
   b. Each of the three alternative plans must be analyzed for advantages and disadvantages with regards to their form and function, land uses, impact on city infrastructure, and may be required to include an architectural site yield study to include potential square footages, parking ratios, and overall land coverage.
   c. An illustrative example of each of the alternative plans that will designate acreages for each use, zoning, densities, circulation patterns, and parking areas
   d. A summary report for the three alternative plans.

3. Final Plan:
   a. A detailed final plan will be based on the analysis of the three alternative plans, and should include the information that was compiled from the first two phases into phase III. This is intended to establish the best approach to developing the site.

Ultimately, when all this work and a large amount of other reporting and compliance requirements were completed, there was still no guarantee that the SLD would ever auction the state trust land to private developers. Then, as now, the SDL had the right to keep the land in the state trust if the results of the planning study or work was deemed unacceptable to uphold the intended uses of state trust lands. The role of state trust lands and lands departments will be explored further in Chapter 2.

With regards to Civano, the Tuscon Solar Village, no planning permit would be issued until the most basic question was answered: which piece of Arizona State trust land would be most suitable for a village-scale development? So, in August 1982 the Arizona Solar Energy Commission (ASEC) Land Committee was formed with the purpose of identifying an appropriate parcel. Miller began discussing his innovative idea—efficient housing experiments built on state lands could become an example of what building science could achieve. The ASEC also served to establish guidelines for the use of state lands intended to become home to an energy-efficient housing experiment—then dubbed the Tucson Solar Village (TSV). In doing so, the ASEC citizen volunteers and staff wrote the originating description of their ideas about what a high-performance community would be—a list of traits which ultimately evolved well beyond the basic goals of saving energy and water. Participants in the Solar Energy Commission Land Committee Meeting, August 24, 1982, worked toward defining the following general goals for the project:

- To improve the quality of life of desert dwellers
- To build new housing that the average citizen can afford to buy
- To construct buildings that will have low yearly energy bills while providing a comfortable environment
- To grow an energy conscious community integrating solar design, food production, and an overall reduction in use of non-renewable resources in an effort to approach self-sufficiency.

The Land Committee members realized that an integrated approach to community design would be needed to achieve improvements on a systems-wide level. They were planning for the community to include mixed-density housing, commercial and industrial facilities, a community center, recreational facilities, community agricultural systems, and a waste treatment and recycling center. Solar access and building performance standards would be required. Cogeneration facilities and community-wide solar photovoltaic (PV) energy systems would be encouraged as well. In the 1980s this was a unique undertaking. There were few completely built example projects to help inform the decision making process.

- Seaside is often given credit as the first modern example of new urbanism in the United States. It is an unincorporated master-planned community on the Florida panhandle in Walton County, roughly midway between Fort Walton Beach and Panama City. It was founded by builder and developer Robert Davis in 1979 on land that he had inherited from his grandfather. The town plan was designed by new urbanist architects Andrés Duany and Elizabeth Plater-Zyberk.

- Kentlands is a 352-acre development located in Gaithersburg, Maryland. Kentlands was developed in the late 1980s and also designed by the husband and wife team of architects Andres Duany and Elizabeth Plater-Zyberk. Kentlands is often credited with being the quintessential new urbanist development with approximately 1700 different types of dwellings and over 3000 people in residence.
- Village Homes, located in Davis California, is perhaps the most pertinent early example, combining many aspects of new urbanism *and* building energy performance concerns into one package. This very advanced community was envisioned by Architect Mike Corbett and his wife, Judy. They purchased 60 acres in Davis where 220 homes, 40 apartments, and approximately 17 small businesses now stand in clusters surrounded by open space. The open spaces include trailed common lands with fruit and nut trees watered by the collection of rain in berms and swales which minimize the need for storm drainage.

All of these projects are early examples of the changing landscape of urban development and represent the cutting edge in design. One can see aspects of earlier projects in Civano, but Civano is still an early innovator, especially in terms of residential and commercial zoning and building code advancement.

After considering locations around the state, the Land Committee narrowed the selection down to five candidate parcels in or near the city of Tucson. They analyzed the parcels to determine which jurisdiction they were in (city or county), how many uses the parcel lent itself to, the short-term (two year) development horizon, and the proximity of existing infrastructure and transportation corridors.

The parcel that was chosen lent itself to multiple land uses while conforming to existing area zoning patterns. The parcel was available for immediate developments and located on major city and county transportation corridors. The varied topography of the site offered excellent views. Its close proximity to a proposed city golf course bolstered the market potential for the final development and increased the rate of return for the state land during the auction. All main infrastructure (water, sewer, and power) were located close to the parcel and the city portions of the parcel were already partially planned under a previous planning permit, which was anticipated to save considerable time and money moving forward with planning. The primary drawback of the parcel, as viewed by the land committee, was that the portion of the parcel that was located in Pima County jurisdiction had never had a planning process started. This meant that there would be additional costs associated with potential project timeframe extensions and planning requirements—but the other four parcels had significantly more challenges to overcome. And so the seed of Civano had an approved plot of land where it could be planted, 820 acres on the southeast corner of Houghton and Irvington Roads in Tucson, Arizona.

## CONFORMING

Just as the early founders of the Tuscon Solar Village had to grapple with the questions of finding the ideal location for development, most green building programs include the consideration of a proposed project's location and linkages as a primary concern. To better understand the context, it is beneficial to take a wide-scale, more regional look at Tucson—Civano's city. The following was National Oceanic and Atmospheric Administration's (NOAA) official description of Tucson's geology, geography, and weather near the time of Civano's master planning:

> Tucson lies at the foot of the Catalina Mountains, north of the airport. The area within about 15 miles of the airport station is flat or gently rolling, with many dry washes. The soil is sandy, and vegetation is mostly brush, cacti, and small trees. Rugged mountains encircle the valley. The mountains to the north, east, and south rise to over 5000 feet above the airport. The western hills and mountains range from 500 to 4000 feet.
>
> The climate of Tucson is characterized by a long hot season, from April to October. Temperatures above 90° prevail from May through September. Temperatures of 100° or higher average 41 days annually, including 14 days each for June and July, but these extreme temperatures are moderated by the low relative humidity (Yes, there you have it—it's a dry heat). The temperature range is large, averaging 30° or more a day.
>
> More than 50 percent of the annual precipitation falls between July 1 and September 15, and over 20 percent falls from December through March. During the summer, scattered convective or orographic showers and thunderstorms often fill dry washes to overflowing. On occasion, brief, torrential downpours cause destructive flash floods in the Tucson area. Hail rarely occurs in thunderstorms. The December through
>
> March precipitation occurs as prolonged rainstorms that replenish the ground water. During these storms, snow falls on the higher mountains, but snow in Tucson is infrequent, particularly in accumulations exceeding an inch in depth.
>
> From the first of the year, the humidity decreases steadily until the summer thunderstorm season, when it shows a marked increase. From mid-September, the end of the thunderstorm season, the humidity decreases again until late November. Occasionally during the summer, humidities are high enough to produce discomfort, but only for short periods. During the hot season, humidity values sometimes fall below 5 percent.
>
> Tucson lies in the zone receiving more sunshine than any other section of the United States. Cloudless days are commonplace, and average cloudiness is low. Surface winds are generally light, with no major seasonal changes in velocity or direction. Occasional dust-storms occur in areas where the ground has been disturbed. During the spring, winds may blow strongly enough to cause some damage to trees and buildings. Wind velocities and direction are influenced by the surrounding mountains and the general slope of the terrain. Usually local winds tend to be in the southeast quadrant during the night and early morning hours, veering to the northwest during the day. Highest velocities usually occur with winds from the southwest and east to the south.
>
> While dust and haze are frequently visible, their effect on the general clarity of the atmosphere is not great. Visibility is normally high.

Tucson is a place of extremes. It is a sloping valley surrounded by beautifully jagged mountains. It often seems as though Tucson is either hot and dry or hot and flooding wet—but in winter it can drop below freezing. However, there are advantages to a thirty degree day to night temperature swing, gobs of solar energy, occasional downpours, and comfortable winters. Once one gets somewhat used to the heat, and learns when to avoid it, living in the Sonoran Desert becomes increasingly appealing for many. The monsoon season brings on a burst of green plants, toads, flocks, and a much-needed break from the famous dry heat, in exchange for a semi-tropical feel. There are areas within the Tucson region that are veritable oases, like Agua Caliente Park with its lukewarm artesian spring water feeding a lush pond-filled park. Fortunately, that place is preserved because in the desert a place like that could be more valuable than oil or gold.

Or is it? What is the actual value of clean water, in the desert or anywhere else? Clearly, when it comes to water, life is hanging in the balance. Thus, in a sense water is priceless. Yet we put a dollar amount on the gallons of potable water delivered to almost every code-compliant building in an American city. As our knowledge grows, our concerns become more focused. The question of where and how to build becomes less about what is convenient now, and more about understanding long-term ramifications for our actions on the environment and the utility it provides to humanity. This question will be addressed in several instances throughout this book starting with the query of how one organizes, conceptualizes, and views the proposed development of a place.

# Introduction to the Charter of the New Urbanism

The Charter of the New Urbanism (The Charter) helps bolster the Congress for the New Urbanism's goal to "support an American movement to restore urban centers, reconfigure sprawling suburbs, conserve environmental assets, and preserve our built legacy." In doing so, the Charter scrutinizes various scales of development, location, function and connectivity to take home meaning about what effects our development and design choices have on the lives of people. The Charter states that development patterns within metropolitan areas should take place as infill in a manner to protect the fringe and untouched environmental resources, as well as minimize the economic and social impacts of development.

The Charter also states that appropriate new fringe development should be arranged as neighborhoods and districts with integral connectivity to the larger metropolitan area, whereas developments that are not contiguous with the metropolitan area should develop with their own clear urban boundaries, like Civano—a town or village with employment and services available on site. These would not be bedroom communities. The Charter of the New Urbanism is like a regional scale plan book. The Charter is not code language, it is a form of pattern language from which one can extrapolate generalities into specific meaning given a specific context—in this scenario, the development of the 820 acres on the southeast corner of Irvington and Houghton Roads in Tucson, Arizona as a potential area of human habitat expansion. Many of the following sections parallel some of those found in LEED green building systems. For access to LEED guidelines and handbooks go to www.usgbc.org.

## LOCATION, LOCATION, LOCATION

If one knows Tucson and Civano, one knows that the issue of smart location is a sticky topic where Civano is concerned. According to the LEED—Neighborhood Development Pilot Version guide, the smart location requirement is intended to guide development toward existing communities or public transportation infrastructure. The general intent is to limit urban sprawl and reduce the impacts of our car-centric infrastructure. Other benefits from a tighter urban infrastructure come out of limiting vehicular miles—such as the opportunity for more individuals to walk to work, school, parks or the store.

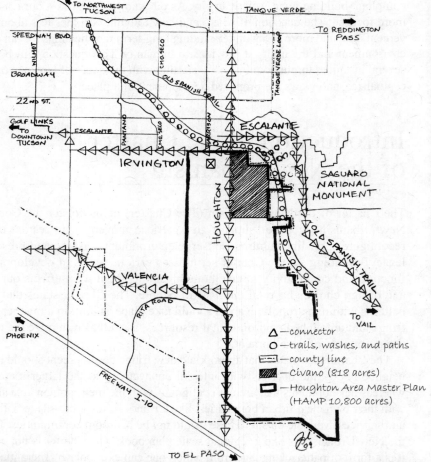

**Figure 1.2** Area map showing Civano, Houghton Area Master Plan (HAMP) and primary travel routes.

Map 1 provides insight into the Civano location debate, showing the 820 acre parcel located far to the southeastern extent of the Tucson city limits. The disconnected aspect of Civano from the rest of the city is clear—what is not generally understood are the reasons why it ended up this way or the developmental context. Civano is the product of a public and private partnership and a process that relied on the allocation of state trust land for planning under public view and the auctioning to developers. This narrowed choices from the beginning and although the chosen site is currently limited in access to transit services and existing shops—it is in a major growth corridor.

Green building programs that define smart location must draw a line in the sand when determining if a project is part of a comprehensive plan within an area to be developed in the future, or if a development is simply adding to urban sprawl. Conceivably, every square foot of land could one day be developed to some extent, but this doesn't justify doing it in an unplanned manner without regard to the future. Thoughts of the future were at the forefront of the planning for Civano and they encompass the primary goals of excellent green building programs.

## PROXIMITY TO WATER AND WASTEWATER INFRASTRUCTURE

Augmenting water and wastewater infrastructure demands very large resource dedication. Even after the initial investment is made, the system will require ongoing maintenance and repairs throughout its service lifespan. Pumping water is one of the most energy consumptive processes undertaken in cities. Locating a project close to installed utilities and waste handling infrastructure will generally come at a smaller environmental cost than remotely located developments. In theory, given similar levels of complexity, a more compact water and wastewater infrastructure system should both cost less upfront to install and cost less to own over its lifetime.

In the case of Civano, gas, reclaimed water, and potable water lines were already installed along the property easement. The nearest public sewer was an 18-inch line located approximately 3400 feet northwest of Houghton and Irvington Roads. Initially, it appeared the line would require the augmentation of roughly 2800 lineal feet of existing sewer, paid for by the land owner, to handle the added capacity required by the rezoning. However, the existing Houghton Corridor already required an even larger sewer to serve development already planned south of Civano—thus rendering the issue not specific to Civano, but development of the region as a whole. During the initial land acquisition and Tucson Solar Village planning, the entire area surrounding the development site was under the South Pantano Area Plan. Today, this planning effort is called HAMP: Houghton Area Master Plan, which constitutes approximately 7 percent of the City of Tucson's jurisdictional land area. Of these nearly 11,000 acres, approximately 7,944 are owned by Arizona State Trust Land—as Civano once was. The state trust land is land managed by the State of Arizona, for the people of Arizona to ultimately benefit from. That means that state trust land is *our* land and how it gets developed, or left natural, affects us all in some way or another—be it a change in neighborhood population densities or the inclusion of a natural park area into the city jurisdiction.

# Framing the Commons

In her book, *Radical Ecology,* Carolyn Merchant describes a cultural movement that is driven to eliminate the types of personal and societal idiosyncrasies that lead to economic forces of production that amass ecological damages to air, water, soil, and biota (including humans). To say that Civano (or any other human built habitat) is just a "drop in the bucket" when it comes to issues of civil infrastructure in the large HAMP region is to suggest a point of view that modern human infrastructure is the best use for that Sonoran Desert land.

Although our society must decide how to maintain and reproduce itself over the long haul, massive paradigm shifts do not come rapidly in the scope of a generation. It seems that people are entering into this second Green Revolution while trying to strike a balance between developing ourselves to death, or succumbing to natural processes that may eventually overwhelm us if we do not learn new technologies. It is almost apropos that the first Green Revolution and agricultural boom fueled by petroleum-based fertilizers, mechanized farm equipment, and hard working people brought much of the food that has helped maintain the amazing population growth on our planet—hastening the need for us to adapt in ways that have spawned the current environmental Green Revolution.

It is a frightening notion that the waxing and waning of our environmental prowess has been greatly tied to the economy. However, regardless of what may be happening in the economy right now, all recent history tells us that unless there is an unprecedented event, growth is coming to our cities in spite of booms and busts in the economy. Civano and HAMP are attempts to manage change rather than just let it happen. Sewers must be bigger, roads widened, homes and businesses built—with the utmost care and planning for the sustainable future we seek. Each pebble we move is an opportunity to do better.

## IMPERILED SPECIES AND ECOLOGICAL COMMUNITIES

The Civano Environmental Resource Report, July 1991, provided a comprehensive description of the vegetation and wildlife found at the Solar Village property. The property is comprised of mostly Sonoran Desert scrub Biome with portions of Sonoran Riparian Scrubland and Sonoran Interior Strand in the northwest portion of the property near the Pantano Wash. Taken as a whole, the dominant vegetation habitats on the property were identified as palo verde-acacia, creosote bush, creosote bush-palo verde, burroweed-mixed shrub, and riparian areas. An extensive survey of Saguaro Cactus communities was conducted and special consideration was given to the Sonoran Desert Tortoise—a state and federal candidate category 2 species.

Solar Village property habitats were grouped into three categories indicating different sensitivity and vegetation levels. The Riparian Pantano Wash region was identified as a class I wildlife habitat—very sensitive to both direct and indirect impacts caused by development, especially changes in hydrological patterns. However, the riparian zones within the Solar Village property had already been made use of—supplying raw materials to supplement the built environment via a sand and gravel operation. Much of these areas had been rendered devoid of naturally occurring vegetation; only occupied by scrubby second growth and low wildlife densities by the time the environmental impact statement was created. Class II and III wildlife habitats tend to be increasingly shrubby, with fewer large plant species and higher tolerance to the impacts of development.

## WETLAND AND WATER BODY CONSERVATION

The nearest things to a wetland on the Civano site are the sensitive Riparian Wildlife Habitats—and they had already been damaged at the site by sand and gravel mining. There were no other high densities of wildlife, species, or unusually high diversity of species in the project area. From a developer's point of view, the property offered few ecological obstacles to building there. A hydrology analysis was conducted for the site which defined the average annual rainfall that could be expected on the site. The report offers useful information regarding where the expected quantities of water would drain off of the site on its way to Pantano Wash. Understanding average and peak flows that could be experienced annually, or during a 100 year storm event, allows for the accurate sizing of municipal drainage systems and the delineation of flood-prone regions.

In some design settings, this may be the extent of one's consideration of the site hydrology. However, even as early as this study, suggestions surfaced regarding the potential of using site runoff to rejuvenate the class I wildlife habitats that had been damaged earlier by pit mining operations—eluding to the system-level thinking that is now beginning to become more widespread within the building industry. Especially in the desert, the management of site drainage is being viewed less as a "nuisance" and more as an opportunity. Many permaculture strategies can be seen in Civano today. The use of berms, swale, gabions, and rainwater collection system retrofits slow and retain rainwater on site.

## USEFUL AGRICULTURAL LAND CONSERVATION

The zoning of the Civano property was originally SR (suburban ranch) which allows for a low density level of residential development with lot sizes of 10 acres or more. For the purposes of the city, the SR zoning label is a flexible designation that can be used to "hold" land until which time more specific uses are established that suit the development plan. The LEED—Neighborhood Development program considers agricultural land conservation a prerequisite of green building, intended to preserve irreplaceable agricultural resources by protecting prime and unique farmland and forest lands from development.

Although the Civano site was not designated for agricultural use, nor extremely well suited for it, the building of infrastructure displaces native vegetation and wildlife just the same. J.R. McNiel's *Something New Under the Sun* describes the history of land use and agriculture. The overarching trend has been the rapid reshaping of natural land cover such as forest, woodland, grassland and pasture into cropland—cropland accounting for over 15 percent of land cover by 1990. Approximately 10 percent of this transformation has occurred in just the last 160 years. As a matter of perspective, it took nearly 10,000 years for human-kind to transform the first 5 percent of ground cover into croplands. In terms of our development practice, it is necessary to protect fertile and unique croplands, but speaking of sustainability honestly requires that we *at least acknowledge* the fundamental challenge in the quest for sustainability is human overpopulation.

## FLOODPLAIN AVOIDANCE

LEED—Neighborhood Development includes this prerequisite "to protect life and property, promote open space and habitat conservation, and enhance water quality and natural hydrological systems." Most of the requirements of the LEED—Neighborhood Development system can be met in multiple ways. In the case of the Civano site, nearly 23 percent is located within identified flood-prone areas—mostly the areas in or around the Pantano Wash. Although parts of the site are located within the 100-year floodplain, only portions of the site that are not in the 100-year floodplain were planned for development. Consequentially, much of this land that is essentially unbuildable has become some of the community's attractive, trailed open spaces.

Furthermore, locating development away from a watershed or flood zone serves at least two protective measures. One, it protects private investments, public infrastructure, and precious life from being washed away in a flood. Two, the watershed itself is less affected by construction activities and site runoff from paved regions can have a chance to be slowed and naturally filtered before entering the waterways. This minimizes erosion and contamination of natural waterways. More about site drainage and permaculture will be covered in Chapter 8.

## BROWNFIELD REDEVELOPMENT

The development of brownfields encourages the reuse of land where development is complicated by environmental contamination. This has the effect of reducing pressure on undeveloped land. Civano is not located on a brownfield site and would not have attained these points had it been a LEED—Neighborhood Development project. The Civano property was chosen using a different set of decisive factors than most private-only developments due to the public/private partnership. Any incentives that a "green" building program can offer to entice developers into reusing some of our currently unwanted land will have the effect of saving pristine land from development.

Although there are disadvantages to using "second hand" property, there are also many potential advantages. Brownfields, which are generally tracts that have already been modified, damaged, or otherwise developed, tend to be nearer the urban core than undeveloped land and may already have some of the basic infrastructure on site or nearby. With creative and proactive planning, the positives can outweigh the negatives. Civano's class I wildlife zones had been damaged by a gravel operation. Although this is not a contaminated site, the impact to the damaged areas was a more acceptable cost than developing near healthier sites.

## HIGH PRIORITY BROWNFIELD REDEVELOPMENT

The development of high-priority brownfields not only reduces pressure on undeveloped land by reusing land that has been contaminated, it also encourages the cleanup of contamination in areas targeted for redevelopment. Developing high priority brownfield land will often require a remediation stage wherein steps are taken to remove and properly dispose of contamination, waste, or old landfill before the site can be prepared for building.

Although remediation can be a costly process, much of this cost can often be offset by the comparably lower cost of high priority brownfield property. When coupled with incentives from green-building programs emerging in municipalities around the country, there is hope that many of the abandoned properties of our built environment can be brought back to life and serve a more useful purpose—while postponing the development of virgin lands.

## PREFERRED LOCATIONS

The LEED—Neighborhood Development system, NAHB's Model Green Home Building Guidelines, and most other formulated green building programs encourage development within existing communities and developed places to reduce multiple environmental perturbations associated with sprawl. A well chosen location will help minimize development pressures beyond the limits of already existing developed lands. By locating new development close to existing development, fewer natural and financial resources will be required for the construction and maintenance of the additional infrastructure.

Civano was located on the periphery of the City of Tucson and this is one of the greatest criticisms of the project. The planners of Civano did not have the same choices a solely private developer would, nor was there an 820-acre parcel of state trust land available for infill within the City of Tucson. There are potential benefits in placing a project like Civano in a future growth corridor, even if it is somewhat disconnected from the urban fabric at first. It often suits large retailers such as Home Depot to place a retail location on the periphery of growth because they can acquire the appropriately sized piece of property and be well established by the time the growth hits. Now that the Houghton growth corridor is just ripening for development, Civano is already becoming well established and may serve as the model for growth for an additional 12,000 acres (approximately).

## REDUCED OIL-FUELED AUTOMOBILE DEPENDENCE

The highly negative effects our transportation infrastructure has on the environment and our culture are now publicly known, even if not fully understood. Although the Eisenhower Highway System was a boon of its time and still provides an amazing utility for widespread cross-country travel, the urban/highway interface has become troublesome at best. At worst, highways and roads have perforated and divided the urban fabric to a detrimental extent—often carving devastating scars across our cities and towns where the human scale is lost, and simple forms of transportation such as walking and bicycling become onerous, useless, or impossible. The toll of the automobile infrastructure on human health and safety is not easily quantifiable, but it is clear that there are a large number of deaths directly caused by auto accidents alone. What about the casualties or health concerns of a sedentary lifestyle, air pollution, global warming, and resource depletion?

Development in locations that exhibit superior performance in providing transportation choices or otherwise reducing motor vehicle use is a tricky business. In the context of environmentalism, the car and its derivations are quite often demonized. However, it is little known that the wide adoption of the automobile played a role in reducing immediate strain on our forests. Train infrastructure before 1920 was growing at such a rate that the nation's forests could not sustain the demand for railroad ties. The automobile also displaced another form of environmental perturbation that created urban pollution. One horse, the pre-car analogue, needed as much land to eat and live healthfully as was required for eight people. Big cities had to clean and dispose thousands of tons of horse dung and carcasses each year which were a source of smell and disease (McNeill, 2000).

All of these untold benefits of the auto compound the fact that the modern day car is a phenomenal tool that has also brought us quick responding ambulances and fire engines, day-trips across 500 miles made easy, and a motivating factor for technology innovation. So, although the automobile has been damaging, it is one step on our path of trying to improve. We should remember these lessons as we make choices with the enormous investment of our transportation infrastructure. Sometimes, we are only deciding between the lesser of two evils.

## BICYCLE INFRASTRUCTURE

There has been one mode of personal transportation that has persisted, time honored and greatly unchanged in basic form and function for many years. The bicycle has been around since at least the mid-1800s and has played a role in transportation, sport, and leisure. The purpose of a well-contrived bicycle network is to promote bicycling—a supremely efficient transportation technology. Civano is located along one major thoroughfare and has many inner connective trails, but riding a bicycle away from the site toward Tucson's center requires a brave soul indeed. Although Civano itself is very safe for riders and appropriately scaled for bicycle travel, unfortunately, there is not yet a bicycle network that connects Civano to the outside world. In 2007, the City of Tucson approved a large road improvement plan that includes widening Houghton Road. This addition will include bicycle lanes, but won't be completed for several years.

## FANTASY ISLAND

Due west of Civano, not even 500 yards as the bird flies, is Fantasy Island. An impromptu bicyclist haven of trails maintained by community members, Fantasy Island draws people from across the city and has been in threat of being closed on several occasions. The story told by Tucson City Council Member Shirley Scott reveals some of the innovative ideas about commuter villages that would greatly enhance Civano's bicycle connectivity to the region, and some of the resistance for such change. The topic is "Fantasy Island," as a trail heaven for bicyclists and a potential boon for developers:

> **McDonald:** At the site across Houghton from Civano, there's the Fantasy Island Park. I'm a huge cyclist. I use it often. Do you know anything about that site?
>
> **Scott:** Yes. As you know, Fantasy Island is a user-designed and built, off road, bicycling trail complex, maintained by the members of the public who use it for recreation. At one time Fantasy Island was in jeopardy of being bulldozed. I helped the bicycling community put together a task force, whose mission was to help draft a plan for future residential development of the area to include and preserve Fantasy Island.
>
> An imaginative developer conceived a plan to develop Fantasy Island with a residential component. By including the adjacent landfill site property in his plan, he hoped to provide a bicycling attraction and amenity with commercial/residential/retail options along Irvington Road. He proposed to use the "borrow pit"(the source for fill-dirt for layering the landfill) as the site for an underground parking garage. He planned to capture the methane from the landfill and use it for some of the buildings that would be associated with the parking garage. His plan would have created retail, restaurant, coffee shop, motel, and other commercial venues on the expanded site.
>
> The Fantasy Island complex is located ideally near Saguaro National Monument East, Houghton Road, and Irvington Road, all of which have existing bike lanes. Just across the street on another landfill there is a BMX track. As you can see, there already exists a synergy among all of these bicycle uses. Fantasy Island is famous worldwide and already attracts bicycle enthusiasts from many countries. If this developer's plan were to be adopted, then the Fantasy Island hub could conceivably serve as a future residential community for bicycling enthusiasts, complemented by commercial development.
>
> A state-of-the-art velodrome would be a perfect fit to complete the picture. There is room on the landfill site to construct such a velodrome. It was hoped the improvements to the two sites would have created a bicycle hub of Tucson. Since part of the land in question is owned by the State of Arizona, the state land department would have to approve the plan.
>
> I made it very clear, I think, to the state land department, that we definitely want Fantasy Island to stay where it is, that you can make it into a commercially, residentially viable project, and that all of the plans and the people involved in this are professionals.

There are a number of new modes of transportation that fit or expand on the bicycle scale that will be further mentioned in Chapter 3. But the question of how to get around efficiently and safely is a key ingredient to building smarter, more sustainable communities. A community resource like Fantasy Island can be the inspiration piece for making that happen, if community resistance to change can be swayed in favor of such developments. A place like the Fantasy Island nearby a place like Civano enhances the value of the bicycle and walking-scale commuter pattern developed in both places.

## LOCAL (WALKING/BICYCLE SCALE) HOUSING AND JOBS

Balanced communities with diverse uses and employment opportunities reduce energy consumption. Pollution from motor vehicles is reduced by providing opportunities for shorter vehicle trips and/or use of alternative modes of transportation. Proximity is the basic element that creates a pedestrian environment where one can usually take shorter range transportation options such as walking or bicycling. On the scale of the whole city, Civano's closest major employment center is the Davis Monthan Air Force Base. However, the base is not close enough to practically travel to by foot and the first mile or more of road leaving Civano is void of safe bicycle lanes. Other large employment centers such as the university technology park, the university, hospitals, and the Tucson International Airport are located further away. Only the rare individual will do anything other than drive or carpool to commute between work and Civano.

Although The Solar Village was located on the periphery of the city, it was located in a corridor with a high growth projection. Any green building program must draw a line between what *is* and what *may be* when deciding what constitutes a smart location. After all, it is truly unknown if some major change event will slow or stop growth along the Houghton corridor. If this were to happen, Civano would forever be located on the periphery of the City of Tucson. If all goes as predicted, Civano will be located within what is being called a potential "second city" within Tucson.

Even if growth in Tucson halts, perceived as a worst case scenario for our economy, plans have been made to allow for many jobs within Civano itself. Because Civano has been built on a more human scale, on-site jobs intrinsically mean fewer vehicular miles that need to be driven by regular commuters. Theoretically, this means that many of the individuals who work and live within Civano may have fewer expenses during difficult economic times than those who live in a traditional urban setting.

## SCHOOL PROXIMITY

When schools are located near the homes of students, it encourages students to walk to school, and consequentially reduces the number of vehicular miles driven within the community. This can help increase public health and well being through physical activity and reduced vehicle-caused particulate pollution. Furthermore, a regular event wherein community members are "out on the town" can promote community interaction and engagement—even if that "event" is as simple as parents walking with their children to school.

There is a K-5 charter school located within Civano and a middle school is planned as part of the future community development. In 2008, the Civano charter school was named "The Greenest Grade School in America" in a contest that valued the children's participation in writing about what it means to be "green." The school may be one of the greenest new school buildings in America as well. The $50,000 prize that was awarded to the school will be used to expand its facility to include a straw bale kitchen. As the school expands, so will the children's education about high-performance buildings and communities.

## Protect Cultural Resources

Archeological resources cannot be replaced if they are destroyed during the development process. Although one may argue that archeology has little to do with protecting the environment, saving these resources opens the door to learning about our past mistakes and triumphs as a people. The Civano Environmental Resource Report included archival investigations to determine the potential for finding prehistoric and/or historic sites within the Civano property. As it turns out, the Civano site was home to Hohokam people probably sometime around 1000 AD and home to a 1920's homestead.

In order to make sure the archeological knowledge was not lost, it was recommended to instrument map and photograph the area and artifacts in detail. The Hohokam findings were to be archeologically tested to understand function and chronological placement and mechanical subsurface testing was performed to determine if there were buried deposits. Sites rich in archeology may take years to process and are expensive to develop. Although archeology may not be directly associated with environmental health, it must be considered as part of land choice for development. However, knowing about what came before is important to understanding the natural environment and the role our cultures and technology have played in shaping it.

## Fires will burn—a perspective on energy and carbon cycles

Most modern energy production technology is based on one of the earliest forces humans learned to focus—fire. Our ancestors spent untold generations learning about the incandescence and radiant heat resulting from various oxidation processes. The exothermic chemical reaction can be harnessed to incite convection, warm bodies, cook food, vaporize water, smelt ore…it is endlessly useful in creating technology and processing raw materials. The use of fire is one of the things that may be as quintessentially human now as the day we became the first known animal to proactively manage fire's heat. Still, most electric bulbs that have replaced a hearth or candle are still lit by the heat of a fire burning somewhere.

Over the years people have been developing a keener sense of how different fuels burn uniquely, achieve different levels of heat, and create different chemical reactions. We've learned how elemental fuels, hydrogen and oxygen, can be mixed to launch a rocket or create a bomb. We can harness the unmatchable nuclear reaction. Metaphorically, all these processes are somewhat similar. They are processes that release radiation energy during some kind of chemical or nuclear exchange that was started by exciting the material to a threshold where it begins to take on different properties and break into smaller parts that then recombine with elements readily available in the environment. The energy that is given off during the reaction sustains the reaction until the fuel is spent. Regardless of what we've learned in the last few hundred years, most fuels used by people have been organically formed and have been burned with the addition of atmospheric oxygen and heat—a basic fire. Some of these processes are considered carbon neutral, and some are not.

This planet supports life that extracts carbon from the environment and concentrates it into bodies of wood, flesh, shell, or other various plant and animal materials. Our human collective of some 6.5 billion flesh-and-bone bodies are in fact a piece of the carbon cycle. Atmospheric carbon is soluble and gets absorbed into our oceans where sea life builds coral reefs and shells out of a carbonate material. As shellfish die, they leave behind their carbonic shells which collect and deposit on the ocean floor or get ground into sand on beaches and buried in the earth—just one of nature's carbon sinks (also can be thought of as a carbon "pump" which moves carbon from one place to another).

We are carbon "pumps" too, and on several levels. When biological matter is digested in our stomachs, the symbiotic bacterial strains in our digestive systems create a hydrocarbon, methane, as one of several byproducts. Methane is both a hydrocarbon and a potent greenhouse gas. When burned, Methane releases carbon dioxide as one of its smoke products, just like the fossil fuels gasoline or propane do (another way we "pump" carbon). Animal bodies and many plants hold hydrocarbons—just think about how a piece of meat can catch fire and burn violently if left unmonitored on the grill or how vegetable oil can flash into a destructive fire. So, one may ask what the difference is, then, if carbon is all around and mixed up from different sources and is sequestered by various means, and why do we worry about which fuel we use?

In the context of a modern world, "which fuel to use" is the wrong question to ask. More to the point would be, "which diverse palate of fuels should we use that will yield the greatest level of health and longevity for the greatest number of people." Sadly, our current knowledge of physics tells us that we simply cannot use fuels without creating some kind of physical change or reaction in the world. Equal and opposite reactions, conservation of energy and mass—we simply must try and choose the least damaging fuels to help maintain our ways of life without inadvertently hurting ourselves further.

Biofuels are viable so long as replacement of the fuel crops is feasible, and attained, without creating any kind of world food shortage or excessive strain on the environment from intensive farming. Wood has been the most long-utilized biofuel by humans. Many pre-industrial streets were lit from the fuel energy of tallow from whales—a biofuel. The first diesel vehicles were powered by peanut oil. Some cars have been retrofitted to run on wood gasification processes whereby the combustible gasses are extracted from wood in a high heat, low oxygen environment. Those gasses are then burned in an internal combustion engine. So these cars literally ran on wood smoke. But the exhaust from all of these sources is still loaded with carbon dioxide.

Imagine if all our cars ran on wood. The forests would have been long gone by now—or we would have developed a much more pervasive worldwide lumber-fuel industry that could perhaps be as much of an environmental perturbation as is our current dependence on oil. This is why energy diversity is important.

High potency petroleum and natural gas come from eons-dead carbon based life and propane is a byproduct of refining these biologically-derived fuels. These are considered fossil fuels. What makes the use of fossil fuels so very enticing is their relatively huge power density. A barrel of oil contains approximately 5.8 million Btu's worth of energy.

Algae is suggested to be the planet's longest lived organism and is given credit for transforming enough of earth's carbon dioxide atmosphere to oxygen to pave the road to the rest of life as we know it. Algae is also said to be the predominant source of petroleum-based oil, having been deposited in layers generation after generation, encapsulated for thousands of years undergoing a refining process of geological proportions. Logically, algae also has merits to become the world's leading biofuel crop—potentially yielding upwards of 45 times more gallons of oil per acre than soy beans.

The various biofuels that are becoming more commonplace today, such as ethanol, biodiesel, wood pellets, and syngas, are all considered potentially carbon neutral because the carbon released in burning can be sequestered in a short period of time by new plant growth. Fossil fuel, still part of the carbon cycle and potentially replaceable, cannot be sequestered in a short geological time span via the same process that created the oil, coal, and natural gas in the first place. Therefore, these are considered non-carbon neutral fuels. No matter what kind of fuel we use, the end efficiency is only as good as the technology we use to focus it. One Btu of energy is one Btu whether it came from a dung-fueled fire in a highland Peru community or the afterburner of a warplane. The length of carbon cycle for each fuel is very different.

Returning to this day, and Civano—remember that the question of efficiency is a question of *how much* we use. The question of energy and the environment is a question of *what kind* we use. Thus, a photovoltaic powered home is not the same thing as an efficient home any more than a biodiesel truck is more efficient than its petroleum counterpart. Assuming equal comfort levels are attained, a well-designed, well-insulated, and well-operated building is more efficient than one that is not; regardless of what fuel is used. Solar power, however, is king for the day.

## A spectrum of solar power

Perhaps the only energy source that people have utilized longer than fire is the nuclear reaction of our sun. This 'fire' is burning approximately 8.3 light-minutes away, or 93,000 miles on average, and has been the fire to light the way for algae and all earthly living things. The amount of raw solar radiation energy that hits earth's outer atmosphere can be expressed as approximately $1.37 kW/m^2$ (kilowatts per square meter). By the time it filters down through the atmosphere and gets absorbed or reflected, the amount of energy that reaches the ground varies, but is approximately $.34 kW/m^2$.

When we use the sun directly and passively, like to heat a barrel of water, the energy is literally free and fluctuates around a fairly constant average value. One can just roll the barrel out into the sun, and it will heat up to some extent so long as the sun is out. This is passive solar heating. There are several different levels of passive solar energy use. Painting the barrel black would help it absorb solar radiation better, but this is still considered a passive use of solar, since the barrel is just sitting in the sun.

Other approaches manipulate solar radiation to focus from a larger collection area (lenses, parabolic, or concave mirrors), to a smaller collection point. The quintessential example is a magnifying glass. It is simply amazing that a very small magnifier can focus a small area of sunlight on to a tiny point that reaches combustion temperatures. These types of technologies are used passively, for the most part.

Some active water heating systems circulate coolant between a heat exchanger in a water tank and a solar heat collector on the roof. More about solar technologies will be explored in Chapter 7, but the general understanding about solar power as a fuel source is that using it instead of burning the typical earth-based fuel offsets part of the human-driven carbon "pump."

## SOLAR RADIATION, WIND AND RAINFALL

If one is considering building a high-performance structure or community, the question of how to increase energy and water efficiency is paramount. After the building envelope efficiency is maximized, diminishing a building's carbon footprint is often best served by the use of solar or wind energy. Rainwater harvesting can supplement landscape water requirements that can be a significant percentage of a household's water use. The Civano site was studied to determine the average wind speed is 8.7 mph, it gets a peak of 350 kBtu's per square foot of solar radiation and can expect 11.2 to 13.3 inches of rainfall during any given year. Knowing these numbers for a prospective building site is important for sizing rainwater collection systems and solar arrays. Wind power was never used in Civano. The noise created by conventional turbines was not considered appropriate in a dense urban housing development setting.

## BACK TO THE PLAN

There are many plans and thoughts that go into a carefully chosen site for a (high performance) development. Deciding to develop a site is the same as deciding to not leave it natural. That is why green building programs require or suggest the use of land that is either already altered by development or close to the existing urban infrastructure. As can be seen by the driving force of urbanism—the rapid move of people away from the country and into the city, we are far from being able to stop the growth in our cities. That leaves us with the option of guiding growth as smartly as possible, with as much planning for our future as possible.

Aside from questions of location and linkages, there are many ecological considerations. What will be the effect of development on the habitat? Are there endangered species here? Is this an animal migration corridor? What is the site hydrology? As answers are attained from questions about location and ecological impact, only then does a bigger picture about the land being developed come into shape. After planners and urban designers become familiar with the property, its intricacies, and its limitations, they can begin thinking about the possibility of actually planning the development itself.

Wayne Moody led the first true planning effort for Civano as a specific place with distinctly human form and function. When he was brought on board the project, he recognized its potential and considered his experience with Civano personally transformational.

**Moody:** One of the things that turned the light on for me about the role of planning in a community like this is that after every single talk, no matter what kind of group it was; people would come up afterwards, like almost drooling, and say, "When are you going to start building? Is there a place I can sign up now? Do you want a deposit?" They were so excited about the concepts of the plan. And of course we couldn't sell yet, you know; it was still owned by the state. There was no private developer or anything like that.

But that was so moving for me to see the marketability potential of a community that really addressed the deep human aspirations that people were expressing. And as part of planning process, I tried to develop that in—we had 20 or 30 community meetings where people would come and express what they wanted to see in the plan. As part of the planning process, it was a new process for the state. As for the state land, urban trust land, Governor Babbitt at the time had pushed through a new state planning act which required that before the state could sell land to a developer, it had to be planned and zoned in the community that it was a part of, as opposed to previously when developers would see a piece of property and say, "Yeah, I'll buy that." It's grazing rights, right? That's the value. So they'd just buy it for the grazing value. And then they'd do the planning and zoning and reap the benefits of the increased value.

Well, Governor Babbitt changed all that. And so not only did he require the planning and zoning be done in the community before [the land] got sold, but you had to do a very thorough environmental analysis and you had to do a very thorough marketability study to determine if the land was suitable to do anything. If it's not suitable, you stop the process. All right, if you ran into a big environmental problem, the process would stop. Or if there was no market for what you wanted to do, the process would stop.

So we had to do those two things first. And once those were approved by the state, then we could get on with the planning.

# THE PUBLIC/PRIVATE PARTNERSHIP

## A Balancing Act

> *Politics is democracy's way of handling public business.*
> *We won't get the type of country in the kind of world we want*
> *unless people take part in the public's business.*
>
> —DAVID BROWER, Savior of Arizona's Grand Canyon

American citizens have the right and ability to affect land development planning which determines the course of the built environment. However, often we find immediate lifestyle demands require the ordinary citizen's time and attention, overshadowing long-term concerns for the environment. The future of our lands, public works, businesses, and homes is a significant matter.

Consider the phenomenon of the sellers' market. When credit flows and economic times are good for homebuyers, homes can be built and sold almost flippantly. The term "flipping a home" takes on a whole new meaning in the environmental context of a development feeding frenzy. Concerns for the long-term environmental costs of unrestrained development often become secondary to the immediate competitive drive to make profits while the market is hot.

So it is clear that there are forces of economics that influence land-planning decisions, not based on the greatest good for the greatest number of people, but on maximizing financial gain or market position. Large corporations are often heavily involved with land development. Incorporating separates the holdings and liabilities of the organization from that of the individuals responsible for managing the organization. Although it is true that there are countless responsible corporate entities, corporate managers are frequently bound by mission statements that do not necessarily represent an even-handed public versus private point of view.

Therefore, citizens must take action by involving themselves with the politics of place because land development is big business dominated at its upper echelons by large corporations. It is up to local citizens to provide some oversight and be active in putting a human face on the future of our own environment. Citizens are provided with nearly everything they need to undertake this project. The main ingredient not provided by law is activism. But honestly, who wants to sit through building code committee meetings, right? How boring is that? According to Al Nichols, there are definitely less-than-dull moments.

**Figure 2.1**
City of Tuscon logo used during the planning of Civano, emphasizing a focus on community.

Jason Laros's interview with Al Nichols:

**Laros:** Al, how did you get into the Metropolitan Energy Commission (MEC)?

**Nichols:** How I got into MEC... I guess it must have started earlier. It would have been about 1991, or maybe 92 when I got drafted into the building code committee. They meet once a month and work on changes to the building code. About that time, the building code introduced for the first time an energy code. They called it the Model Energy Code. That would have been the 1992 code. We reviewed it and they came out with the next version of it in about 1995, and that was about the time I got on the energy commission.

There was a call from John Guthrie. He used to be with Tucson Gas and Electric back in the days when it was a monopoly, before they broke up and became two separate companies: Tucson Electric Power and Southwest Gas. He had a natural gas powered air conditioner. There are very few and he had his for years and years. He would religiously go out and change the oil, the spark plugs—just like a car. And he would brag about how cheap his air conditioning was because it ran on natural gas. He was probably the only person in town who had one. It eventually broke down enough to where he couldn't fix it anymore and he had to go out and buy a heat pump. He was a fun guy. At any rate, one day, John told me he was retiring off the energy commission, and asked if I would take his place. I said, "What energy commission? I never heard of it." He said, "Oh, you only have to come to a meeting once a month" and so I said, "Alright."

So I got appointed to the energy commission. I had to sign a loyalty oath. I got a card I had to carry with me that said I was a commissioner on this energy commission. They told me the meetings were at seven o'clock. So I show up downtown at TEP headquarters for my first meeting and the guard wouldn't let me in the building. I said, "Look, I'm a commissioner. Here's my card and my letter, we're meeting here at seven." He said, "Nope" and showed me his meeting book. Nothing was in it.

So the next day I call up and explain that I was there at seven, figuring they must have moved the meeting or something happened. I had wasted a whole hour or two. At any rate, during the conversation something came up about "morning" and I found out that they met at seven o'clock in the morning, not evening! Oh, man! I had missed the first meeting by exactly 12 hours. Who meets at seven o'clock in the morning?

So the next month I show up and I have the worst cold, flu—whatever. And I was just barely able to sit there in the back trying to figure out what is going on. Everybody was discussing lively subjects, particularly this one about the Tucson Solar Village. By the third meeting I was feeling better and I began to realize what was going on. This was a 2500 home development on 820 acres. I was adding this all up ... even at $150,000 per home and 2500 homes, that is three hundred and seventy five million dollars and things weren't yet really getting done! So this was serious.

I then got appointed to a sub-committee looking into the Tucson Solar Village and discussions there were even more lively! Reports and tons and tons of paper were flying around and people were trying to figure out what to do. So it turns out that I eventually chaired a smaller sub-committee to figure out what would become the Sustainable Energy Standard.

We had only just passed the Model Energy Code through the building code committee that was hard fought. Many of the folks in the construction industry who were sitting on the building code committee were sure this new energy code was going to raise the cost of housing and make it unaffordable. So that was a terrible fight. It came down to the last vote and my colleague, the swing vote. He thought we should just let the marketplace determine the code.

But there was a council member in the audience and she was just amazed at how ...I won't say violently, but it was loud—how loudly it was discussed. Heavily opinionated. And she had never seen anything like it. And at the very last minute, my colleague decided he would change his vote if we called it the "Sustainable Energy Standard" and not the "Civano Code"...

...It was so dynamic that it took me a day and a half to convince the council person we had actually won by one vote! So, we had an energy code and I was merrily happy with that.

Not every development policy meeting will be as contentious or exciting as these, but they are all open to the public in case one wishes to observe democracy in action. Open meeting laws pave the way for public oversight. It is valuable to read such a law at least once and know of their existence. They protect a basic right for people to know what decisions are being made where they live.

## THE STATE OPEN PUBLIC MEETING LAW—TITLE 38, CHAPTER 3, ARTICLE 3.1

Arizona's open meeting law of 1962 requires all public meetings will be made open to the public and must be preceded by publicized agendas which contain enough information to inform the public of the matters to be discussed and then decided. In this context, the definition of "public body" means the legislature, all boards and commissions of this state or political subdivisions, all multi-member governing bodies of departments, agencies, institutions, and instrumentalities of the state or political subdivisions, including without limitation all corporations and other instrumentalities whose boards of directors are appointed or elected by the state or political subdivision. Public body includes all quasi-judicial bodies and all standing, special, or advisory committees or subcommittees of, or appointed by, the public body. The following text is not extremely riveting, quoted directly from Arizona's Tile 38 governing public officers and employees; but sharing local public meeting laws is the first step toward empowering the public to participate in the politics of place:

**A** All meetings of any public body shall be public meetings and all persons so desiring shall be permitted to attend and listen to the deliberations and proceedings. All legal action of public bodies shall occur during a public meeting.

**B** All public bodies shall provide for the taking of written minutes or a recording of all their meetings, including executive sessions. For meetings other than executive sessions, such minutes or recordings shall include, but not be limited to:

**1** The date, time, and place of the meeting.

**2** The members of the public body recorded as either present or absent.

**3** A general description of the matters considered.

**4** An accurate description of all legal actions proposed, discussed, or taken, and the names of members who propose each motion. The minutes shall also include the names of the persons, as given, making statements or presenting material to the public body and a reference to the legal action about which they made statements or presented material.

**C** Minutes of executive sessions shall include items set forth in subsection B, paragraphs 1, 2, and 3 of this section, an accurate description of all instructions given pursuant to section 38-431.03, subsection A, paragraphs 4, 5, and 7 and such other matters as may be deemed appropriate by the public body.

**D** The minutes or a recording of a meeting shall be available for public inspection three working days after the meeting except as otherwise specifically provided by this article.

**E** A public body of a city or town with a population of more than 2500 persons shall:

**1** Within three working days after a meeting, except for subcommittees and advisory committees, post on its Internet website, if applicable, either:

**a** A statement describing the legal actions taken by the public body of the city or town during the meeting.

**b** Any recording of the meeting.

**2** Within two working days following approval of the minutes, post approved minutes of city or town council meetings on its Internet website, if applicable, except as otherwise specifically provided by this article.

**3** Within 10 working days after a subcommittee or advisory committee meeting, post on its Internet website, if applicable, either:

**a** A statement describing legal action, if any.

**b** A recording of the meeting.

**F**  All or any part of a public meeting of a public body may be recorded by any person in attendance by means of a tape recorder or camera or any other means of sonic reproduction, provided that there is no active interference with the conduct of the meeting.

**G**  The secretary of state for state public bodies, the city or town clerk for municipal public bodies, and the county clerk for all other local public bodies shall distribute open meeting law materials prepared and approved by the attorney general to a person elected or appointed to a public body prior to the day that person takes office.

**H**  A public body may make an open call to the public during a public meeting, subject to reasonable time, place, and manner restrictions, to allow individuals to address the public body on any issue within the jurisdiction of the public body. At the conclusion of an open call to the public, individual members of the public body may respond to criticism made by those who have addressed the public body, may ask staff to review a matter, or may ask that a matter be put on a future agenda. However, members of the public body shall not discuss or take legal action on matters raised during an open call to the public unless the matters are properly noticed for discussion and legal action.

**I**  A member of a public body shall not knowingly direct any staff member to communicate in violation of this article.

## LAND DEVELOPMENT MEETINGS

Special advisory committees appointed by public bodies take a large role in zoning, building code, and planning efforts. These matters are also decided in open public meetings conducted by the responsible public entity itself. Representatives of the public, such as a city council, will hold meetings to ratify or deny the informed decisions or counsel of these advisory committees. Thus, zoning and building code development is subject to a public review process both during the planning stages and when decisions are being made. In addition, the resulting zoning maps are made available to the public.

In the case of Arizona's Pima County, initial zoning for the county had been completed by 1953. So, one may ask, "If the zoning was decided over half a century ago, how can I have any role in the regional planning?" The basic answer to this question highlights a citizen's opportunity to influence the built environment because as times change, land-use planning changes. When governments, developers, or business owners desire to change the traditional zoning designations to better suit their wants and growth plans, part of the process becomes public. A concerned citizen ready to be active can watch for public notices of applications for variances and rezoning-all of which are subject to public review. If only the builders or developers directly involved with the project show up to public meetings to educate the officials who ultimately pass laws, then there is minimal direct public oversight to the process—if any.

Excerpts from Brian McDonald's interview with Tucson City Councilwoman Shirley Scott can shed some light on how a person serving on a public body may experience a project such as Civano as it becomes part of the public purview. Some of the subjects and people mentioned in their discussion will be elaborated upon in later chapters.

**Scott:** My name is Shirley Scott, and I was first elected to the Tucson City Council in 1995. In December, the background information and first discussions of the Solar Village came before the council. Because the identified location was ideally suited to the proposed use, I supported it strongly. I am proud to have played a role in helping to create the Solar Village with a 4/3 vote, in favor.

**McDonald:** When did you first become aware of the Sustainable Energy Standard, the SES?

**Scott:** I'm not sure I could pinpoint a date when it became a standard. But in terms of the Solar Village project, its requirements, and the new way of looking at things, more sustainable than ever, I thought it was extremely intriguing. I learned that the State of Arizona had already designated a suitable site for such a development. I had also heard there were at least a couple of people interested in purchasing the property in order to create a solar village, I believed that it was a good probability on the horizon and was looking forward to seeing how it evolved. After our vote, two gentlemen stepped forward and purchased the property and immediately started to plan it as such and put up the first building for the Global Solar Company.

**McDonald:** Have you met Kevin Kelly and David Case? What role did they play in this, aside from the land purchase?

**Scott:** Yes. I have. They came by to meet me and told me how interested they were in making this a reality. Finally, the Solar Village project was now out of the idealistic, philosophical realm and was moving into the real world. They came by to tell me how they were going to make this work. They told me about the restrictions and how they were going to plan for the infrastructure and how it would evolve from streetscapes, to the kinds of housing, what those housing units would have in them, and how uniquely the overall development would be built out.

Before the houses were built, before the people were there, before Global Solar was there, I was asked to go to Australia to talk about Civano to an environmental group there. All I had to show at the international conference was pictures of the desert and the cacti. So I was really just talking about what was to be, with all of the environmental assets that would be on this piece of property. The international community that attended this particular convention was so captivated with the idea that they were eager to get the literature out of the box that I brought with me. It engendered huge interest. It was the busiest workshop they had in the entire conference. The attendees believed they were looking at the equivalent of a new Olympic Village, which usually means a living demonstration of the newest architecture, innovations, and standards being used.

When I talked about this particular project, this became so interesting to the people present that even those who couldn't stay for my workshop stopped by and literally took all of the literature I brought with me, begged for the website address, and said they hoped to hear more about it and that they would keep in touch.

At the time I made my presentation, which was in the '90s, it was still cutting edge for a community, a city, to do this. In a PowerPoint presentation (PowerPoint was also a new technology at that time) I said, "The City of Tucson has been there, we've done it, we even have the T-shirts." And I had a T-shirt on at the time.

The name "Civano" came afterward. The original name was Tuscon Solar Village. The environmental group that sponsored me [In Australia] was the International Council on Local Environmental Initiatives (ICLEI). Funding for my participation in the event came from the U.S. Department of Energy.

Live, work, learn, play was the guiding principle for a sustainable community. If you live in a place like a "Civano" you do not have to leave your neighborhood to get most of your daily needs met. There's even a small school there.

**McDonald:** As part of my research, I've come to understand the very delicate play between developers, builders, realtors, and home buyers. Do you have any thoughts on that?

**Scott:** Well, I think it was fortunate that in its initial stages, when people were buying brand-new homes, there was a very good market for the Civano model. Civano has experienced great success in the resale market as well. I think that speaks loudly on behalf of the people who were involved or engaged. Now, many years later, people are very energy conscious. Nationwide people are investigating alternative fuels and energy efficiencies. Lessons learned at Civano point the way to the future, illustrating what's already been established and has successfully worked.

**McDonald:** Do you recall when the city adopted the amended SES as standard for all city buildings?

**Scott:** Not the exact date, but when we were talking about sustainable energy and energy efficiencies, there was a rule passed by the city council at the time that said all city council members' offices shall be located in their wards, so that constituents could have easy access to them. None of the buildings you see in this area were here. City staff and I discussed what kind of ward office would be appropriate.

I suggested that since the federal government was offering a rebate on solar installations that we should stop talking about and start constructing a solar powered municipal building. So in the end, the Ward 4 Council office became the first municipal building ever built by the City of Tucson that has a photovoltaic array on the roof and is solar powered.

The Ward 4 office is "net metering" (producing more power than we consume). This building is oriented north/south; it has no windows east/west. It has lights in here that are called "green lights"; each of these light fixtures is calibrated separately and differently. They are independent of one another and do not necessarily all come on at the same time. They respond to heat; not movement.

We have no light switches in most of the rooms. There is only one room in this building that has a light switch so that the lights can be dimmed for audio-visual presentations. The light fixtures in there hold energy-efficient bulbs.

We have water saving devices here. We also have outside canvas fixtures that look like sails, which on the northern side, I'm told by the architect, directs the sunlight in or out, based on the sun's angle in the sky, to direct heat or radiation away from the building as needed.

We also have several zones in this small building. Any unoccupied zone goes dark until someone enters it. We also have some solar tubes for lighting in certain areas. So the building is about as energy efficient a building as could have been built in 1999.

**McDonald:** You have double-paned windows?

**Scott:** Yes, argon filled. Moreover, the city is now incorporating solar and other alternative and sustainable features in all city buildings they remodel or retrofit. I believe that they have replaced all of the old bulbs in all of the city buildings with "green lights."

**McDonald:** From what I understand, you've always been a proponent of Civano; you've always voted very positively toward initiatives that involved it. Do you recall Fannie Mae's involvement with Civano through the American Communities Fund?

**Scott:** Yes. Fannie Mae did take over the development of Civano. I'm not sure about the details, because I don't know the private dealings that go on between the partners and Fannie Mae; that's not part of what I do. What goes on within the borders of the Civano property is pretty much up to the individuals involved. And the City of Tucson was not directly involved in that, you know. I'm just glad that it's as successful as it is and that there's still a demand in this area to consider very favorably such a planned development using all of the most up-to-date, cutting edge energy efficiencies that are available. Our local utilities advertise that if you buy a particular house they will guarantee the amount it will cost you to use their electricity or gas. That's a selling point to the customer. Civano residents have established a strong track record of using energy-efficient appliances and practices. I think that's a positive trend that sets a new standard. And I'm very happy about that.

Five of us on the council, and I think it's just five, now have alternative fuel vehicles. They are the Prius hybrid cars that the City of Tucson is buying as part of the fleet. The police department cannot use them. The City of Tucson was one of the first cities in the country to build an alternative fuels refueling station for its bus fleet. It was available to other agencies, including the school districts and the University of Arizona fleets. We have a pipeline that collects our landfill gases and pipes it directly into our local electric utility. So I think as a community we are doing very well on those levels. We are also probably the single most water conservative community in the state of Arizona.

**McDonald:** Is that recent?

**Scott:** I once got a request from ICLEI to write a short article filling a 3-inch column in their international newsletter talking about how the City of Tucson manages water conservation. I directed them to the water department. Instead of only using a small 3-inch column in their newsletter, they devoted an entire issue to the water conservation efforts that City of Tucson residents have been practicing for the last 20 years or more.

The City of Tucson and we who live here are very conscious of energy efficiency; we're very sustainability-oriented; we are very accepting of practical conservation alternatives that might be available. Since we live in an arid land, it all makes sense to us to do the right thing.

However, one of the difficulties for Civano was that a lot of the technologies are still very costly. And the ability to realize a reasonable return for your investment takes awhile. But ultimately sensible investment does pay for itself.

The challenge is taking the concepts that have been tried in Civano and then transferring them fully to the broader market. To conserve household utilities, i.e., water, electricity, natural gas, and wastewater through improved design reduces costs long-term. So it's always market driven.

Trees and other vegetation around a house can shade and decorate that property, as well as produce edible fruit. Civano also sets a high standard for recycling of yard and household waste.

Councilwoman Shirley Scott was a strong supporter of the Civano project and her specific knowledge about the issues was impressive, especially for an individual expected to have a broad-based knowledge of her ward and the surrounding city. Not every decision maker will have these same strengths. Feedback and participation from public groups and citizens helps strengthen leaders' roles regarding the topics which are most important to us, including the fate of the built environment. Al Nichols talks more in his interview with Jason Laros about how building code development happens on a local level:

One is not only appointed to a committee, the committee tries to fill specific disciplines that are needed to make building science decisions. Sometimes people want to be on the committee, but if nobody is there to fill positions on the committee, the committee members will solicit individuals to become members. So it is not just a political appointment.

**Laros:** How much influence do these committees have?

**Nichols:** We don't have any power, per se. We have the power of recommendation over a select group of presumed professionals. But it is a great process. Seldom, if ever, are our recommendations challenged. For example, there are just two of us on the subcommittee of the energy code committee of the building code committee and we meet in the basement of Development Services. We decide the changes we are going to make to the energy code and then we bring that to the building code committee and it moves on up. So just the two of us write the energy code for Tucson, just because nobody else wants to participate. So if you want it your way, you've got to show up!

**Laros:** How is the MEC different from a code committee?

**Nichols :** Well, in both cases they are citizens' committees comprised of people who have taken a loyalty oath, carry a card, and serve at the favor of the city council or Pima County supervisors. So appointees follow all the same rules as elected officials. They can get in trouble if they don't follow all the rules or do anything unseemly, like somehow make a profit or steal money. I even got in a little trouble once for voting to pay a commissioner to help build the group's website because she was a web master and would do it for half-price. They pointed out that was highly irregular, we had to go out and hire a web master for twice the price and half the service. But they let it slide.

So even with the best intentions, that is the sort of thing you cannot do as a commissioner. One is literally a public servant, even though they are not getting paid. Well, you do get free parking.

**Laros :** These are public, open door meetings?

**Nichols :** Yes, building code and energy commission meetings are always public meetings. The agenda has to be published 24 hours ahead. The commissioners cannot advance anything that is not on the agenda, it is just like a mayor and council meeting. Quite often, contractors sit in for their own purposes, especially if they think there is something on the agenda that will affect their business. We always give them a chance to voice their opinion.

**Laros :** What is the process of developing local building codes?

**Nichols :** In general, what we're doing in the code committee is taking the code, which comes out every three years, and reviewing it to make sure it is compatible with our region. Most codes are not region-specific. They are becoming more region-specific. But at first they were not. So both the building and energy code committees make amendments to the international codes. At one time, the local code considered Pima County as all one region, one code. But Pima County ranges in altitude from about 2000 feet to 9000 feet and has at least two very distinct climatic regions. For the code, we decided to base the regions on altitude; above 4000 feet and below 4000 feet. So especially the energy code, for example, had to be amended to consider both the heating and cooling climates we have in our region so the formulas would work. Then we have to go to DOE and get them to change their ResCheck and ComCheck verification software to reflect the heating degree days and cooling degree days that Pima County has adopted to reflect the actual local regional conditions. So now there are three different standards— the above 4000 altitude standard, the below 4000 feet, and the Sustainable Energy Standard.

The code committees are specifically made up of a diverse group. We always have architects, mechanical, electrical, and structural engineers, somebody from the general public, and staff. Staff would be somebody from the jurisdiction having authority. The Development Services Director or one of his subordinates shows up at our meetings for observation. They are not allowed to vote, but they are allowed to keep us on track on what we can and cannot do in Development Services. So, that's the first step. The code committee findings and decisions go with staff recommendations to the city manager and the city manager puts them on the agenda for the city council and the representatives for the county board of supervisors—and it usually gets passed without much discussion.

That's the neat part about these committees. They are literally made up of mostly experts with no particular agenda so the City Council has no real reason to deny any of the changes we make. And we can always go back at any time if we need to make changes, if the staff decides some of the changes we made were unreasonable. The staff ultimately has the chance to correct anything.

As stated by Al Nichols and required by Arizona law, building code development is also part of a public process. Building code is developed by volunteer committees comprised mostly of professionals active in the building industry. Although a concerned citizen may not possess technical expertise about a specific code matter, usually provided by experts in their given field, each person notices different things about their environment and may offer insights unforeseen by the professionals and officials most often involved with code and planning matters. Most members of building code committees are members by invitation, but building code meetings are open to the public for observation and/or participation.

During the planning of Civano, times were different than during this writing. The idea that a significant number of mainstream developers would willingly take the "risk" of spending more money to create an environmentally friendly masterplanned community on their own volition was very unlikely. Today, the divide is narrowing between the idea of promoting zoning and code that are concerned with public health and safety versus zoning and building code that will protect our interests today *and* in the future.

Contemporary public awareness is creating a growing demand for socially and environmentally responsible products—including our cars, food, housing, and even whole neighborhoods. This makes the incentives motivating how developers and consumers assess the costs and benefits of investing in high-performance design entirely different. In a nutshell, the financial risks associated with "green" building are diminishing day by day, lessening the need for the dynamics of a public and private partnership to raise the bar on sustainable development—as was the process for Civano.

Now, developers are more apt to go beyond the status quo to provide products that will reach a growing target audience of home buyers interested in energy efficiency and sustainability. Sadly, some still only offer token actions that will support soundbyte marketing statements that have little credence, if any. Indeed, "greenwashing" has flooded our televisions, radios, and store isles; grazing on our emergent environmental concerns and our conscious like too many sheep on a field. Unchecked or misunderstood claims about the "greenness" of products, processes, and services are among the most egregious dangers to the advances of eco-conscience industries. What is at stake is the public's education about the effects of their consumer choices on the commonwealth of the work done for us by nature—the cleaning of air and water, the carbon cycle, the fisheries, the forests, and all the pieces in between. To lose these would be a catastrophe—one of which we cannot "greenwash" our way out.

# Reframing the Tragedy

*We abuse land because we regard it as a commodity belonging to us. When we see land as a community to which we belong, we may begin to use it with love and respect.*

—ALDO LEOPOLD, Quoted in Stewart L. Udall's *The Quiet Crisis* (1963)

The commons is our shared environment. The tragedy of the commons, touched upon in Chapter 1, is when we ruin the services provided by the environment for each other. The traditional depiction of "the commons" uses the metaphor of a shared field where all the shepherds in the region come to graze their individual flocks. Nobody owns the field—nobody polices it. The field and the nutrition it provides for the sheep trickle down to the people who have domesticated them. The grasses and water from the field serve as a common natural resource to the local society—just as our oceans, atmosphere, and the carbon cycle provide functions from which all life on earth benefits. This analogy could easily apply to builders developing land to sell homes.

This view of the commons creates a dynamic whereby an individual's action (such as adding animals to his herd) is chosen based on the perceived results of a nearly 100 percent chance of personal gain. This perceived gain is combined with a nearly 100 percent chance that the costs of the individual's action will be shared by everyone using the commons (Hardin, 1968). Expanding the metaphor, suppose a doubling of flock numbers constitutes the amount of overgrazing that would end the natural production of the field due to the combined effects of consumption, trampling, and waste products. If there are 100 shepherds and one decides to double his herd, he takes on 100 percent of the benefit of that action and only 1/100th of the consequences.

The consequences of a 1 percent increase in overgrazing may not amount to enough hardship for the other 99 individuals to notice. The increased production of the one individual, however, will most likely be noticed by others. This may create a trend whereby 10 more shepherds decide to double their herds and the shared damage from overgrazing theoretically jumps from 1 percent for each shepherd to 10 percent each. At this point, one can imagine how the incentives of each group change. There are those who are noticeably producing double the sheep while the other 90 percent of the people are now trying to maintain the same number of animals on land that is approximately 10 percent overgrazed. So, the math of the 'free' commons never yields a balanced equation.

The rational shepherd must make some kind of choice to either quickly double his own herd now to make use of what is left, effectively doubling his benefit while raising the risk that the land will become altogether unproductive, or try and maintain status quo and hope he can make due on land with diminishing resources with the hope of being more sustainable. Carried through to its ultimate conclusion, the perverse incentives created by this treatment of the commons results in a natural tendency for each shepherd to maximize his or her perceived personal gain at the expense of the group-to the eventual destruction of the commons and loss to all. Therein lies the *tragedy*.

In 1962 Rachel Carson's aforementioned book, *Silent Spring,* opened America's eyes far beyond any one small field to the fact that chemicals and pesticides people were using in their private lands were getting out (into the commons), concentrating up the food chain, and killing keystone species—widely affecting ecosystems beyond the fields where these technologies were used. With this awakening, we began to see that as populations expand and our technological practices yield more environmental perturbations, the commons is vastly more complicated than a simple field.

Private land ownership does not remove land from the commons. The commons are in everything we touch, from the grain on our plates to the plates themselves and the chairs we sit on. The basic truth is that we've taken our resources and piped them all through our homes and places of work and play. We no longer go "graze" the open land in few numbers, but digest the Earth's natural resources from the comfort of our human-built habitat. But now, perhaps we are beginning to realize the consequences of bringing the commons home with us-inexorably bound to the *tragedy*.

Anybody involved with developing land, building, or maintaining aspects of our built environment, is undoubtedly aware of this shift in perception and some of the ramifications it is having on the development industry. It is important to remember that sometimes when the public at large is outraged by the misuse of non-renewable resources, our new view of the commons encourages us to take action in hope of correcting the misuse. When this happens, somebody must pay-in other words, the "field" is now being policed. But what is a logical punishment for those found to have misused the commons?

# Oil Overcharge Funds

In the Internet age when it seems everything is at our fingertips, we the people are happy to abide by fair-to-middling prices and even shortly lagging availability. If what we need becomes difficult to attain or we feel the threat of it becoming scarce, we will often do what we need to get it-even if what we want or how we get it is bad for us, or the earth, in the long run. It is no surprise that oil is among the most threatening commodities of this nature because our current global economy is nearly completely dependent on the compact energy stored within that amazing hydrocarbon stuff. Oil has become a pump that moves food, commodities, people, and all other matter of environmentally-derived goods that have become ubiquitous with the most influential human ways of life.

So when our business model for getting inexpensive food, for example, relies on using cheaper foreign labor to keep costs down-the need to transport the food longer distances to where it will be consumed becomes an integrated component of the whole process. Thus, when oil gets expensive, food often gets expensive. If in our minds oil is equal in importance to food (in a sense oil *is* food if there is no local food supply) we may naturally decide to go after oil, whatever the risk.

Large organizations are subject to the benefits and risks associated with managing enormous quantities of a precious resource. In the 17th century, Spain fell from being a dominating world power into a vast recession due to lack of economic planning, having spent the wealth brought from the Americas without concurrently developing the basis for a post new-world exploration economy. During the time of this writing, a food shortage in widening parts of the world has catalyzed rioting and starvation in many less affluent nations, like Haiti. These are but two examples of the risks associated with the management of highly sought-after (or *needed*) world commodities. Large energy companies have also run the gambit of managing natural resources with extraordinary world-wide demand, and at times have paid a price for mismanaging the same.

## CIVANO GETS A PETROL-FUNDED BOOST

Several times, major energy companies have been suspected in violation of price control regulations and put under federal lawsuits. If found in violation, the companies were required to pay into a federally managed escrow account intended to be distributed to benefit the primary victims of oil overcharges perpetrated by the organizations. These funds are commonly called petroleum violation escrow (PVE) funds or oil overcharge funds. Those who received oil overcharge funds were end users of petroleum products such as motor vehicle operators, electric utility customers, and home heating oil customers. The rules guiding the allocation of the overcharge funds were conceived to reach several general program outcomes, energy conservation, or other energy-related benefits with federal approval and reporting required.

In 1983, Exxon Corporation was found liable by a U.S. District Court for overcharges on domestic crude oil. Several years later and after much litigation, the court's decision was upheld and Exxon was directed to pay 2.1 billion dollars into escrow accounts to be managed by the U.S. Department of Energy (DOE). The money was to be divided between the affected states and used exclusively in grant programs for programs as mentioned above-energy efficiency projects.

Coming back to Tucson—in September 1985 representatives of Pima County met with local builder organizations (John Wesley Miller, Southern Arizona Home Builders Association), the state land department, the governor's office, the University of Arizona, and other public and private entities to discuss the feasibility of using the urban land transfer process to create a project that would showcase an energy efficient and environmentally conscientious development. Their deliberations resulted in the Tucson-Pima Metropolitan Energy Commission's (MEC) application to the state requesting $250,000 to initiate planning of the showcase community earlier envisioned by John Wesley Miller.

Thus, the unfolding story of Civano was given a chance when in 1986 Bruce Babbitt, Governor of Arizona, dedicated approximately $210,000 from Arizona's Exxon PVE funds to the Metropolitan Energy Commission (MEC) to support the planning of *Tucson's Solar Village (TSV)*. There was a planning process program championed by David Elwood-an individual with 60 years experience in all aspects of land development, planning, and landscape architecture:

> **Elwood:** I'm a professional planner who's been involved for over 60 years. I worked here at the Pima County planning department right after I got back from Korea. I've been an urban planner, land planner, environmental planner, landscape architect, and all of those things. And I've been frustrated all my life with planning and how ineffectual it has been. It has created a mess. I thought this was an excellent opportunity to address the issue, crystallize the problem, and envision something that would really be beneficial and serve as a prototype for the twenty-first century for land development.
>
> I worked up a planning process program with Karen Heidel, then city energy office manager, and submitted it to her. This basically got the money flowing. We decided we needed a project manager so we went into the process of interviewing a project manager. The whole thing came under the Metropolitan Energy Commission. That is when we hired Will Orr. That was part of the planning process of what we needed to do. It was to get this thing into a planning program that would serve the community, and be focused on saving energy-because that was stipulated as a requirement for use of the oil overcharge restitution money.

The MEC sponsored and directed the project and in 1986, and as part of David's efforts to provide a planning program, Will Orr was hired as project manager by MEC. Orr was hired as county employee and paid with federal funds from the Department of Energy (DOE) and housed in city offices-demonstrating the integrated support for Civano. Orr, an experienced green designer from British Columbia, in turn contracted the planning to Wayne Moody-but they were provided vision and mission statements from MEC to help guide them.

> **Elwood:** This was the meat of the whole thing, why we wanted to do it and what we wanted to achieve. This is how it started. To me, part of the planning process is defining the problem statement, the vision, mission, causes, and effects. All that has to be defined up front so you know what you're going after.

The Civano/Tucson Solar Village problem statement presented to Wayne Moody's design team was, "To plan and develop a solar and environmental showcase that will transform and guide Tucson's suburban areas into the twenty-first century by responding to and creating imaginative solutions to mounting social, economic, and environmental problems in the form of an implementable master plan."

The counterpart vision statement was, "To create a compact pedestrian/biking community with shopping, services, and employment interwoven into a variety of housing types and open space; and to create a sense of place and visual order that radiates identity meaning and belonging." The vision also sought to "Accommodate the dynamic change occurring in social and economic demographic groups."

The mission statement born from the planning program stated, "Create a more self-reliant and sustainable pedestrian and biking oriented, suburban community designed to achieve energy and cost efficiency, resource conservation and environmental stability-all being in a visually pleasing setting that encourages human exchange and community responsibility."

As planning progressed, it was directed by The Civano Oversight Committee and guided by community members-architects, engineers, planners, builders, educators, the local utility companies, and many others. The planning unfolded in a participatory process that included considerations for community resource efficiency, site design, water, solar, waste reduction, and transportation. The integrated design process used then is becoming ever more ubiquitous with the action: sustainable development. Civano is a model in its own right for a multi-sourced, multi-level public and private partnership.

The developmental course of Civano took a track beyond the vision of a solar village and moved into the realms of new urbanism or neo-traditional design, with solar. The design considered may peoples' point of view, benefitted from grassroots support, and flourished through a glorious contention that lives on today. Perhaps, contention is merely a greenhouse gas to ambition. From this point on participation in Civano grew to involve many people from many walks of life. One individual, Bob Cook, possesses expansive knowledge about the politics of place and the environmental concerns associated with the development of human habitat. His words are most appropriate to establish certain perspectives on the times leading up to, and out of, Civano.

Excerpts of Brian McDonald's interview with Bob Cook:

**McDonald:** Please tell me your earliest recollection of Civano.

**Cook:** Back in 1989 I was working with a group of people that just launched a new 501c3 called Nest. More than 200 people attended our introductory meeting that started this whole organization off. And our primary focus at that time, the late '80s, was exploring the concept of community and building community. Some people were interested in intentional communities, and some people were interested in intensifying our sense of community within Tucson itself, our neighborhoods, our interest areas, and so on.

And we read in the newspaper that the Metropolitan Energy Commission, the City of Tucson, and Pima County had received a planning grant from the state as a part of the oil recharge funds, the Exxon overcharge funds. So anyway, we heard in the paper that this project was being launched and it was a consortium of state, county, city, and the energy commission. The metropolitan energy commission was given sort of oversight-comprehensive oversight to-to the process. We had some meetings at Nest with regard to this opportunity because, you know, here was a public-private effort developing. And the idea was to come up with a vision and a set of goals for what would be Tucson's first planned, model sustainable community.

We were very interested in sustainability at that point since some of us had come out of the appropriate technology movement of the '70s. In the mid-'70s, I was working on the East Coast in Philadelphia and New Haven. In 1974 I was involved in a research project where we looked at all sources of energy including all the renewables as well as conventional fossil fuel sources, and nuclear energy sources and technologies. At that time, I think we had examined around 20. That work resulted in the book called "Energy Earth and Everyone," published in 1975. Also, another important invent in the 1970's was the limits to growth study. I went to a major conference in '75 called "Limits to Growth" and I met people like E.F. Schumacher, Lester Brown, Amory Lovins, Dennis and Donella Meadows, Hazel Henderson, and all the people that were talking about shifting to a lower growth world with fewer environmental impacts.

So the whole idea of appropriate technology, earth-friendlier technology, was really hot then. Well, it was hot in the circles that I was in. I mean, it was a small movement but there was quite a bit of interest. The solar energy market was booming as a result of some of the tax credits at that time. Jimmy Carter's administration was very aggressive. He had one of his Saturday evening chats to the American public while wearing a sweater, talking about the moral equivalent of war.

The mid-'70s was a period of the Arab oil embargo where the OPEC countries cut off their supply to the Unites States. And there was about a 7 percent curtailment in the flow of oil in the United States. And we had an enormous reaction. We had 20 percent interest rates, 10 percent unemployment, and almost a decade-long recession. The '70s was a difficult time for the country. And energy was viewed as a big part of that.

So there was a lot of consciousness around-awareness about energy and energy technology. It also turned out that the U.S.'s own oil production peaked in about 1970. So we were becoming more dependent on imports. And you have to remember; at one time the U.S. was the Saudi Arabia of oil production in the world. We had the most reserves and the most production-a major oil producer in the early part of the century.

This interest in energy diminished somewhat starting at the beginning of the Reagan administration. In fact, a lot of tax credits for wind and solar were eliminated, and it was also a time where OPEC decided to pump more oil and to lower the price of oil down to $10 a barrel to remove all the incentives for development of alternative technologies. And so solar had a tougher time in the '80s to develop. But there were some visionaries in Arizona. One was Carl Hodges-the founder of the environmental research lab. Arizona has had a number of colorful figures. John Yellot was one of them. And this goes back to the '40s and '50s when people were talking about developing solar energy. Since Arizona has so much solar energy, there were scientists throughout those decades that were interested in developing it.

The founder of the University of Arizona Optical Science Department, Aden Meinel, was also a big solar enthusiast in the '50s and '60s. So there was some background in terms of interest in developing the solar industry. And there were discussions in the mid-'80s after this opportunity came up about the recharge funds to look at a larger scale of development, more than just a subdivision, but a community scale development.

## THE TUCSON INSTITUTE FOR SUSTAINABLE COMMUNITIES

The Civano Village Corporation was the original Civano not-for-profit entity. It was later named The Civano Institute and then renamed again as the Tucson Institute for Sustainable Communities. As one of the project stakeholders, the institute was a non-profit organization whose mission was to cultivate vital and sustainable communities to be resource efficient, ecological, affordable, and equitable.

### The First Civano Report – July 1990

A quarterly publication issued by the Civano Institute, the *Civano Report* was a newsletter dedicated to informing the public about the Tucson Solar Village project. It bore the seals of the City of Tucson, Pima County, the Arizona Energy office, MEC the U.S. DOE and was produced in part by the Project Manager, Wilson W. Orr of P&D Technologies. An Excerpt from the first Civano Report reads: "Our goal is to build the *best community possible* by merging Southwest traditions with environmental caring, solar design, proven technologies, economic opportunities, social responsiveness and sound financing." Three primary project objectives were put forth:

- **Environmental Accountability** utilizing energy efficiency with Solar applications, resource conservation and recycling, biking, and walking for internal transportation, water conservation, and preservation of open spaces.
- **Economic Performance** with employment provided for 50% of the residents, primarily in environmentally useful occupations in a business district enhanced with home/studio offices and civic/educational facilities.
- **Social Responsibility** through affordable, handicapped accessible housing and community child/elder care facilities with continuing educational and youth employment opportunities.

The desired result advertised in the first Civano Report of this public and private partnership process was a community that would use less energy, create less waste, cost less to operate and have fewer harmful effects on the environment. Zoning for "The Village" was accepted by Tucson's City Counsel in Oct 1991. Civano's most general form was decided. After zoning, the scale of the community was relatively clear in terms of basic housing and retail densities, open space and permitted uses—but there were still many differing perceptions of what the community would ultimately be. Master and specific plans were still needed. Debate within the design team of professionals, civil servants and the community revolved around the key concerns about marketing and the deeper-reaching goals of Civano as an example of how future development in Tucson and around the world could be done better.

*One of the biggest lessons learned is that the marketing of a new and innovative development idea is an integrated design element of the project, not a stand-alone piece of advertisement.*

Planning efforts and community debate did include the topic of marketing. The team naturally considered who their customers may be and why they would want to live in a place like Civano. By and large, the people involved with the planning were not the ones who ultimately lived in The Village as it neared full build-out. Yet 14 years before this writing, during a community input session, one community member voiced the opinion that presented more as a query, paraphrased here: "Why would we have not defined our market pre-design, rather than while we are picking out final masterplan details?"

The balance between 'forward vision' and 'too forward vision' needed to be struck, an idea that will continue to be developed more throughout the book. What is clearer now than it was then is the question of how far the status quo of an industry can be pushed must not be overlooked. In many ways, Civano is a victim of the very compromise that was necessary for it to go forward. The highest-end niche market would have required a *pure* Civano, with no barrier to sustainability left unsolved. A general market could not tolerate the increased cost of such advanced specialization on such a large scale in Tucson.

There are multiple existing barriers to sustainability—from mechanical to social and natural. One interesting example can be found in a mechanism originally intended to remove barriers to sustainability. A common methodology that has been successful in promoting solar and other 'green' technologies has been the use of tax rebates and credits. However, the government itself owns and operates untold numbers of buildings and they don't pay taxes; which means that one of the most likely long-term building owners to benefit from solar energy must pay more to get it. Similar situations can occur for tribal governments or other entities that do not pay taxes. There are many people attempting to resolve various barriers to sustainability and these topics will be recurring throughout this text.

## PRE-PLANNING

Designers are affected by the very way design requires them to think. Consider needs, consider intended use, consider actual use, consider a constructible design, cost and schedule. That is challenging enough. Then, add in an action-based ethical framework based on ecosystem-level integration of our designs—and we tend to find our feet firmly planted in the future. Design is basically the projection of our vision for something in the future on top of what we have now, be it a pad of dirt or an aluminum billet, and sometimes the vision misses a couple points of integration in between. But that is what designers do and their ability to visualize an end that satisfies all project goals is why they earn the big bucks.

If one were to go back in time and sit as a fly on the wall in the hours and days-long development and planning sessions for Civano, he would hear many of the questions asked and concerns raised that have been proved pertinent in the last decade. As one would hope, a majority of these questions were raised by the public participants and contracted planning team, but every bit as many innovative and progressive ideas were offered by the concerned volunteer citizens involved with Civano's politics of place. Not all the correct decisions were made, but the fact that the questions were asked is what remains important. Now we know some of the answers for next time.

In early 1992 the Civano Master Plan was accepted by the State and one proactive builder submitted an offer to purchase property, which triggered the Trust Land sales process at the state level. Unfortunately, changes in the market and the State Land Office's unfamiliarity with the newer process being used to develop these State Trust Lands delayed progress until 1994 when the city council directed the city manager to *aggressively pursue* the development of Civano. Before long, the master plan, zoning and codes for Civano would be up for auction, along with the 820 acres of State Trust land to which those documents belonged. More about the specifics of the planning and zoning efforts, as well as marketing challenges and solutions, will be covered in greater detail in following chapters.

Although it may seem ironic that in the late 1900's oil companies helped bring life to a more sustainable community, it was actually the logical result of a positive effort to protect the commons, consider the post-oil economy and create positive change for our future. However, the legal mechanisms to make a public and private partnership development like Civano possible were set in place much earlier, towards the earliest years of the United States of America existed as a sovereign nation.

## State Lands Department

The forefront of the land planning efforts that led to the development of Civano started over one century before Arizona was even a state—a true testament to some of the unheralded foresight and vision of early land planners. The Northwest Ordinance of 1787 established the practice of setting aside public lands for educational uses. Arizona did not become a state for more than 120 years after the ratification of the Northwest ordinance, but during that time lands were already being allocated in the territory and needed to be managed, including 60,000 acres eventually given for the University of Arizona Trust (thus the term, land grant university).

All told, over 10,000,000 acres were allocated in Arizona to be held in trust for specific beneficiaries and to be sold or developed in a manner that would put the land to its, highest and best use based on the potential use of each parcel. Enter Mulford Winsor, Cy Byrne and William A. Moody—appointed by the first Arizona Legislature (1912) as the initial three people to direct the Arizona State Land Department. Thus, by one of the earliest acts of law in Arizona, the State Lands Department is the manager of state trust lands and is a direct descendent of one of the earliest land management techniques used by the United States government. However, a few modifications have been made to the Department's increasing range of freedom—one of them being the Urban Lands Act.

# Urban Lands Act

Ninety-six years after its formation, the Arizona State Land Department still manages over 9 million acres of state trust land for 14 beneficiaries and over 90 percent of revenue goes to K-12 schools. The passage of the 1981 Urban Lands Act authorized the Land Department to create a development plan for urban lands to increase their value, pre-exchange or sale. This enabled the state to oversee rezoning and use planning of trust land specifically with beneficiaries in mind. The act also allows the State Land Department to hire contractors to conduct planning services or issue planning permits to private developers, thus opening the doors for the possibility of a project like Civano to bloom (State of Arizona Office of the Auditor General Performance Audit Report to the Arizona legislature By Douglas R. Norton Auditor General Arizona State Land Department April 1997 Report Number 97-6).

The ombudsman (or citizen's aid) overseeing Civano has a point of view that includes the role of the Land Department and the Urban Lands Act with regards to this unique project. The role of an Ombudsman is generally to represent the interests of the public by investigating complaints made by citizens and following through by trying to mediate resolutions.

Excerpts from Brian McDonald interview with Hector F. Martinez: City of Tucson Development Ombudsman

> The Arizona State Land Department was created by the State Legislature to manage the land granted by congress to the state for the benefit of common schools and other beneficiaries The state land department , through the Urban Lands Act, later was also granted authority to work with local jurisdictions to develop development plans and entitle the property through a rezoning process. That is how the City of Tucson in 1991 became involved in the rezoning of the property and the name of the project "Civano, a model sustainable community" evolved.

**MacDonald:** So first the state approved the master development plan and then Mayor and Council rezoned it?

**Martinez:** Yes. It kind of came in that way. But if you look at the approval dates of the actual master development plan, which is right there, it's dated March 5, 1992. That's when the state land commissioner signed it into effect, as it was still owned by the state land department. However, it wasn't until early 1995 that the state and the city aggressively pursued a possible purchaser for the land based on this master development plan.

# Metropolitan Energy Commission

One public entity that was integral to the planning and ongoing oversight of the Civano project is the Metropolitan Energy Commission (MEC). The goal of MEC is to serve "...as a catalyst for the City of Tucson and Pima County to build a more sustainable energy future in the region" (www.tucsonmex.org/index.html, accessed May 21, 2008). Other than managing the $210,000 of oil overcharge funds and the initial planning work for the Tucson Solar Village, MEC developed the Sustainable Energy Standard (SES) which, by zoning, is the energy code in Civano and has been adopted by the City of Tucson for its own buildings.

## IDEAL IDEAS

If the most advanced ideas for Civano had come to be, the development would have become a completely off-grid community with a very large PV power station on site. There would be no cars inside the community, leaving them parked on the periphery of a carefully designed, walkable community with integrated land uses. Ideally, cars would not have even been that necessary because the city bus system would be integrated with the community infrastructure as well.

Local food production would have occurred in fields and a greenhouse that was kept warm in the winters by the large, efficient central plant which would also heat and cool all on site buildings and homes. The goal was that one job would be on site for every two homes and services would be within 10 minutes walk from most homes. Given the 2,500 residential units, achieving these numbers would mean there would have been approximately 1250 jobs. All of this infrastructure would have to have been developed in a way that would maintain 50 percent open space in natural desert condition on the site.

It was believed that Rocking 'K' Ranch, a large community planned further east of Civano at the foot hills of the Rincon Mountains, would benefit Civano by bringing in the high-end market. Even so, southeast Tucson had been increasing residential growth by 16 percent at that time. Several miles away came the large IBM facility, nestled on many acres available there for technology park development—a prime nearby employment center that would provide well-paying jobs and a short commute for Civano residents working there.

The early goals also indicated that total water consumption would have been reduced by 65 percent, energy would have been reduced 75 percent, solid waste reduction would have reached 90 percent, transportation designs would have reduced air pollution 40 percent and total transportation energy would have been reduced by 50 percent. It was known that neo-traditional design techniques were going to add cost to the project from the early stages of concept development. Extenuating ideas for edible landscaping, microclimate effects from highly scrutinized planting and orientation schemes, and zero-water landscaping were meant to mesh with anticipated energy efficient mortgages. Unfortunately, this was not all to happen as expected because of disparate perceptions of what constitutes added value when designing a housing development.

With today's standards for high performance design advancing by large increments, Civano's original goals sound like the next step from now. Fifteen years ago these goals were like trying to jump over the Empire State Building in a single bound, with assurance of neither flight nor safe landing. Just imagine the expressions on the faces of those investors first approached to fund the project. Although it was those who were too forward thinking that ultimately pushed the limits of what could be planned at that time, when it came down to funding what was intended to be a profitable, unsubsidized project, compromise was necessary. More about the funding and business challenges of Civano will be covered throughout the book.

The beginning goals of Civano were measured against the concept of sustainable development-meeting present needs without compromising the ability of future generations to meet their needs. The cooperation of Civano's visionary public and private founders pushed that concept as far as they thought it could go-further than it actually could then. Still, the project employs a holistic and integrated approach to developing the built environment and bears the marks of both progressive thinking and the fear of change. Civano stands today as a model and demonstration place for builders, architects, engineers, planners, students, and citizens alike; because of the lessons learned there, but more importantly because of Civano's successes. These lessons learned and successes will be explored in more detail in future chapters.

# 3

# GUIDING GROWTH
## *Master Planning and Analysis*

> *We are never going to save the rural places or the agricultural places or the wild and scenic places (or the wild species that dwell there) unless we identify the human habitat and then strive to make it so good that humans will voluntarily inhabit it.*
>
> —JAMES HOWARD KUNSTLER

At the root of the debate about achieving increased sustainability in the built environment are the concepts of scale and integrated design. Harkening to Newton's Third Law of Motion, "for every action (force in nature) there is an equal and opposite reaction," in turn with land development, every design or development goal (action), affects aspects of scale and integration. This cause-and-effect relationship is not a newly discovered idea, but comes to the forefront when trying to build places while considering how those places will integrate with their surroundings and what influences they will have on nature and people.

Early concepts of Civano heralded high-level aspirations of off-grid utilities and car-less neighborhoods, but many of these bold ideas were not to be. So much of what has transpired throughout the planning and building of Civano has been the compromise between dreams about new systems in the built environment and practical design choices based on the built environment we are working within already. Some residents of Civano are upset with the results of the project, discouraged that the original vision was not fully met. Their frustration is understandable, but the project may never have been built if some of these compromises were not accepted along the way.

Much of the problem has been that these forward-thinking ideas were too forward and too large-scale in terms of complexity and newness. If Civano had been designed to reject the car completely, it would have become isolated from the broader car-centric city of which Civano is a smaller part. The neighborhood-scale solution to this problem is to locate parking areas away from the neighborhood and let people walk, bike, or perhaps use golf carts between their cars and their homes. But moving cars out of sight would not have changed the fact that Civano homeowners still must have their cars to connect with the greater Tucson region. Removing cars from the interior roads may make for a

**Figure 3.1** In Civano's phase I, narrow streets lead to the community center. *Drawing courtesy of Justin Cupp.*

nicer setting inside the neighborhood, but does little for addressing the large-scale issue of the auto infrastructure.

Giving in to the pressures of an existing developmental pattern that is less beneficial than a known alternative may seem defeatist. An optimist may say that we can take a leap of faith, invest in pushing the car out of our neighborhoods and be hopeful that the larger city-scale transportation system will also adapt accordingly. But developers and home builders cannot afford such financially risky optimism. They need to sell building lots and sell homes or their businesses die. This is one point of integration that is somewhat immovable, and must always be part of a project's consideration wherein the developer is, in fact, in the business to make a profit.

Imagine a developer, in 1990, being asked to be the first in the region to build a residential community where everyone had to walk a distance just to get to their cars when those same people would otherwise happily purchase homes with a garage and driveway. They would immediately wonder if consumers would really opt for the walk and if so, how many people would comprise this segment of the market? When millions of dollars are at stake, radical shifts in design seem a large risk to take when tried-and-true practice traditionally yields sales.

Had the existing city-scale built environment in Tucson included a healthy mass transit system that was extended to Civano with the capacity to serve the population there, perhaps the car-less infrastructure idea would have held up. But this is a key example of a system-level design consideration and reasons why no matter how frustrating it may be, some design ideas are before their time. Part of what is exciting about projects like Civano is that the design process makes people address these new issues, hear multiple innovative ideas to challenge current scale and integration concerns, and thus push forward the best practices consensus toward increased sustainability in land development thinking and design practice.

The following excerpts are from individuals who were involved with the planning of Civano. Only a relative few of the participants are represented here because the process involved hundreds of people from all walks of life from across nearly 25 years, each with their unique points of view and ideas. From the least activist home owner to the most ardent new urbanism advocate, what nearly all of them had in common was the desire to do better than they had been—whether that meant living in a nicer home than they had before, or designing buildings that use less water and energy on a daily basis.

# Excerpt of Paul Rollins' Interview with Wayne Moody

**Moody:** I always viewed my role as the chief planner for Civano in this planning process to get to a master plan, I viewed my role as one who had to make it real yet reach the absolute highest potential that you could possibly get. But the bottom line, because being a planner you see a lot of plans just go on the shelves and never happen, I knew that if it wasn't real then it wasn't going to happen. So that was the big challenge for me. And at the time, remember this is still the late '80s, we were just beginning to see the movement in telecommuting; people working more at home. It wasn't really a fad yet, but you could see it on the horizon, or I could see it on the horizon. Other people couldn't. I said, "Well look, people are going to be telecommuting, so we need to develop that factor into Civano to create the kind of environment that those kinds of people are going to want…"

That was a hard sell. You know, it was a hard sell because people-a lot of people—in the community, the home builders, realtors, and so forth thought this would never sell. You're not going to change the people. You're not going to change the way people work and, you know, view that.

And this was just one aspect. The other aspects that we went through in the planning process, [like the question], "Can you get rid of the car?" So we spent a lot of time, I spent a lot of time doing analyses on getting rid of the car gradually over time before, you know, the energy costs are going to get really high. You know, we have to guess about all of these things. People aren't going to want to drive so much. And if they have the kind of environment in which they like to stay, they'll find a reason to stay around; they won't need the car so much.

I remember doing an economic analysis of one of the schemes to have parking structures at the entrances to Civano. People would drive in, park their car, and have little shuttles or little electric cars or something to get around. So that the whole town, the whole village, would be car free. And one of the thought processes I went through was, "Wow, hey, you could actually create a car rental business, right? You could afford to get rid of one of your cars, just have one car." I was trying to push this, you know. Every weekend you could get a different car. Right? It's a cost to you of $500 or $600 a month with insurance, with the cost of a garage, and maintenance, and gasoline, you know, depreciation of the car is what is really costing you. For that kind of money, I could rent a Corvette one weekend. Then I could rent a limousine. I could rent a four-wheel drive. You know. Wow. Who wouldn't want to do that?

**Rollins:** So it wasn't totally clear how you were going to accomplish these things? You had to work your way through some possible scenarios to arrive at what was working at a marketing level.

**Moody:** Yeah. There was no program for what to do with the land to begin with. So yes. One had to kind of guess well, "What's the world going to be like possibly, how real is that, how could you attract the kinds of people and businesses and so forth that will help make that world happen?" And we did a lot of alternatives as part of the planning process. Part of the process was, as consultants we had to come up with three, at least three, different land use and planning alternatives. And then the Metropolitan Energy Commission would evaluate that.

Now in addition to that, before we even did that, the Metropolitan Energy Commission had its own design charrette, another pot of money. I was involved with it heavily. But MEC controlled that process. I didn't like it because, you know, we already had one design charrette; and we're responsible for coming up with alternatives afterwards. A little awkward.

Anyway, they retained people with the Solar Energy Research Institute, the vice president of Economic Research Associates; a well-known architect who used to be head of California state architecture, a big developer that had experience in doing real development and a couple other people. They brought it all together here, in Tucson. It was one of those weekend design charrettes.

And I—we did it at a time when I had just completed all the environmental analysis and the economic analysis. So we had those resources available, no plan yet. And then we put together students from the University of Arizona, we pulled together a lot of resources from neighborhoods around Tucson. It was a big deal. We rented big rooms at a hotel. And then we had this, you know, the standard charrette that just, you know, went on and on. And what is really interesting about that charrette is that they came up with a plan. And that plan is very similar to how the plans for Civano came out after I was finished with it, very similar indeed.

The other really interesting part, which I found fascinating, you know, as any charrette goes with a lot of people with a lot of ideas and egos and so forth, you run into conflicts. So that the head of the charrette, who was this developer guy, decided he was going to divide us up into two teams. We're going to have a team that works on physical form, okay. And we're going to have a team that works on the kind of neighborhood, community mentality thing, because there were arguments going on, mostly between the more intuitive types and the more physical types, about, you know, what each of those factors, what role they should play.

So they divided up into teams. As it turns out, not planned, everybody on the physical planning team was a male; everybody on the community, you know, intuitive team was female. It was most interesting. So the males went about doing the physical stuff. The females went about doing their intuitive stuff. And they each came up with a plan. And as you can imagine, the physical plan had buildings and land uses and, roads and all that stuff together, halfway system, open space preservation, et cetera. All laid out, you know, really nicely.

The female team developed a flower with five petals. And the five petals actually could fit physically on the land, each petal was a place on the land. I still have that drawing somewhere, it's a great story. Each petal was kind of variegated in color; and as it got closer to the center, it became redder and more alive and so

**Figure 3.2** A conceptual drawing of the five petal land development plan. *Drawing courtesy of Justin Cupp.*

forth. And then coming up from between the petals were little dotted lines, you know, that was like energy flowing out. But each petal was a little neighborhood. And the red part in the center was the center of the neighborhood. And the little dotting lines going between petals were the pedestrian paths and bicycle system to get people out in the nature, as it turns out.

And one of the most interesting things, because it was kind of based on the physical nature of the site itself, because they had environmental analysis, you can superimpose those plans, one on top of each other and kind of see each other's, you know, see the opposite plan. Really fascinating process. I loved it. Totally magic.

The planning process for Civano was held to very high standards. Since the project required a total re-zoning of the state trust land being used, the public was to be fully informed and play an integral role in approving the plan. The public meeting exercises mentioned in Chapter 2 were taken to larger measures than minimally required by law. Herein are the basic benefits yielded from Civano's public and private development partnership: a system of checks and balances was established based more on policy designed to reach a higher goal rather than pure cost-and-demand economics would normally allow. Moody acknowledged the inherent weakness of this relationship as well:

> *And if I had my druthers, I would never, never do that again because it took away a lot of the authority that the developer might have if they're a good developer.*
>
> —WAYNE MOODY

If the end zoning and code requirements for idealistic developments are too imposing, no private developer will purchase the property because private developers are in business to make money, usually not to provide a selfless public service. They most certainly will not take more than a well-calculated risk. If something is very new, the risk is difficult to calculate and a natural tendency is to consider unknown quantities as the worst-case scenario. Thus, a balance needs to be struck between the high performance ideals, pie-in-the-sky high-performance development plans, and what can actually be built and sold for a profit for a project as heady as Civano to move from the drawing boards to reality.

Case in point: later, *Habitat for Humanity* wanted to buy into Civano lots as a developer. But by the time the lots were being sold, the price was already out of reach for that organization. As will be discussed in later chapters, much of this added expense was due to expensive development choices that may not have been necessary to achieve the more idealistic goals of Civano that were let go. Thus, some of the large scope (scale) of innovation did not integrate well with some of the aspects of the project goals; one goal being to provide affordable housing. There have been many starter homes built in Civano, even homes that achieved HUD's definition of affordable housing, but *Habitat* in Civano would have been an ideal match. It would have been an opportunity for a blending of public policy, private technology, and a successful educational model for building affordable housing. Nevertheless, planning for Civano was holistic as possible and addressed very many aspects of scale given the pre-established constraints on the project.

**Moody:** This whole process was actually quite short for planning. It went from the end of 1988 or beginning of '89, to like in April of 1992 with the approval of the plan. So it was a pretty short process, basically two years plus a little more for getting the zoning approvals and support.

So to continue on the thread of what it took to get all the support, we did the public meetings, we would get letters from the neighborhood associations. We had support from the Audubon Society, all the conservation and environmental groups, all the neighborhood groups, and there was a big neighborhood coalition. We got back the home builders group, the chamber of commerce. You know, we tried to cover all sides of the issue. We had letters of support and people that would offer to come and speak at the council meeting and the zoning examiner hearing for this.

So there is a zoning examiner's hearing, which is a requirement of the city, you go through the zoning examiner and then you take it to the city council. The zoning examiner had one public hearing, he said it was the best plan he had ever seen, and he complimented everybody on the planning process. It was just beautiful. Went to the city council a month later, set on 7-zip vote to approve. I mean, it was slam-dunk. And that was because people were behind it, it came from the highest aspirations people had, there was really no developer who was trying to get something done, which is both a blessing and a curse. And we had all the support, you know, that you needed.

So it was beautiful. I mean, it was such a high to be able to have. And then of course all the newspaper articles come out and, you know, praising this and that and everything. Well you know, fast forward over 10 years, and it's Sunset Magazine's best new community. Right? Even with compromises. So that, you know, it kind of tells you it's a different way to do planning. All right. You talk to the people first to find out what they want. And then you try and meet their aspirations.

Civano was the type of project that I think was transformative for everybody who got involved in it. And I think the reason for that is it had such high aspirations for changing the way people lived on the land and lived in community with each other, which I think reaches deep aspirations of people. So yeah, why wouldn't you want to be part of that? Because there's a real good chance you can actually make a difference on this. And so people, you know, with heart, come in and say yeah, I want to be part of that. Of course in the process, you know, you have to meet up with the reality of the situation. And then there are challenges. And so it's like this, you know, trying to make a pearl out of a grain of sand in an oyster. That's what Civano really did for people, myself included.

But it was clearly spiritually a joint effort between the city, the county, the Metropolitan Energy Commission and a bunch of people out there who really wanted to see this thing happen. There was a lot of support and backing for this.

Lesson learned: although there are those that are disappointed that Civano did not reach the pinnacle of sustainability that they felt our current-day technology and design techniques could offer, the project leaders did prove that early involvement (integration) of stakeholders is a key project integration component.

During the early planning of Civano, project leaders were very successful in engaging a wide-scale of people who could voice complaints, express wants, name risks, and collaborate on meeting these challenges. Civano has undisputedly attained high-performance housing on a large scale. Is it for everybody? No. Was it as successful as it could have been? No. So much of what was learned is the true meaning of system-level design and processes. Even some of the very strong successes of stakeholder integration were not complete in terms of scale. There were times when Tucson City Council passed rules affecting the process for permitting of Civano buildings without fully coordinating efforts with development services. This caused problems for some builders. Problems for builders often mean more expensive housing than originally anticipated.

Furthermore, integration and scale goes outside of the front-end planning and construction of the project. All financing and marketing considerations need to be aligned with the type of project being built and sold. Some assumptions disregarding the integration of stakeholder education led Civano to be plagued by image problems and financial losses, the stories of which will be imparted in future chapters via stories of "lessons learned" from the challenges of developing and planning Civano.

## Ian McHarg's *Design With Nature*, 1969

The first Civano Master Plan, spearheaded by Wayne Moody, was inspired by the earlier ecologically-rooted design principals put forth by Ian McHarg. McHarg (1920-2001) was a landscape architect and city planner who grew up in the industrialized city of Glasgow. McHarg gained an early appreciation of the need for cities to better accommodate the qualities of the natural environment. McHarg developed these interests into a method for designers and planners to consider the social value of natural resources on the development chopping-block. At a time before there was a way to store, process, or present large amounts of spatial data, McHarg's social value composite analysis used transparency map overlays to develop a visual representation of the ecological components of a particular geographical location, and the anticipated human-built infrastructure.

McHarg's map layering approach was a powerful tool for relaying a large amount of information about a sight and offering a way of gathering new insights about what a proposed development project would add in terms of social value versus the social value that would be lost due to alterations to the existing ecosystem. Using this methodology makes a developer consider what each layer of the map may be. This practice, in turn, relates to an inventorying of a proposed development site's natural resources, geographical attributes, and understanding of the work the given ecosystem does for the environment which benefits society as a whole.

As it turns out, this process was an influencing precursor to the modern, information-rich tools of Geographic Information Systems (GIS). Wayne Moody used very similar map layering techniques for his analysis and planning of the Civano site.

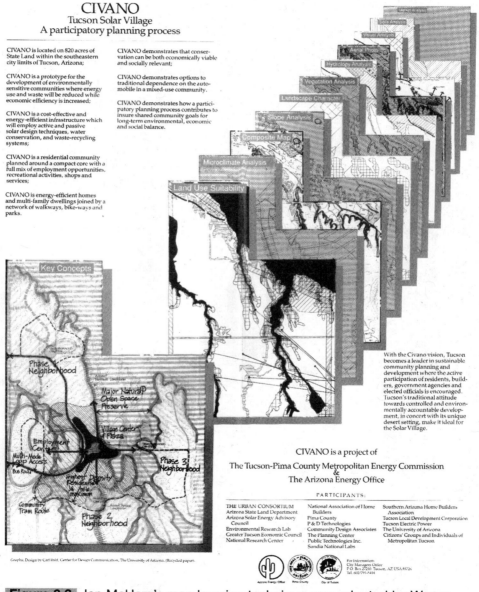

**Figure 3.3** Ian McHarg's map layering technique was adapted by Wayne Moody, resulting in the composite map illustrated here by Carl Rald.

Ultimately, these tools are mirrors of the people using them. GIS, the information-age's version of map overlay analysis, is a mighty tool for managing information. The information can be used with a range of intents from simply developing strategy to get a job done, to determining whether or not a job should even be considered. Wayne Moody, like McHarg, approached design with a sense of balance as much as convention allowed—he used the tools available to him to ascertain if they could design a project that would have equal or greater value to society than the nature preceding it—or if they should not develop that place at all.

**Moody:** There were several things that kind of evolved during the planning process. One of them was the creation of jobs as part of the goals. And that was really—it was not a new urbanism. It didn't come from a new urbanism idea. It came from a sustainability idea. I mean, one of the whole ideas that stimulated the creation of this Tucson Solar Village and the funding was to develop the solar energy industry in Tucson. And to go back, the analyses showed, the follow-up energy analyses of Tucson at that time showed that people in Tucson were spending $1.1 billion a year on energy use: gasoline, electricity, natural gas, airline fuels, and so forth—$1.1 billion a year. And 75 to 80 percent of that would leave the community, it wasn't working in the community. It would go to all the energy companies.

**Figure 3.4** A vision turned into reality. Pima County 2008 aerial photograph with the Civano Phase I drawing inset.

Today, phase I of Civano (Figure 3.4) is described as a "new urbanism" or "neo traditional" community in spite of the fact that its original planners were more naturalists that "new urbanists." But many attributes the old urbanism communities had were due to issues of scale, how far a person could walk to the rail, school, or work rather than where they would park and drive a car. So many of the qualities that are ubiquitous with new urbanism hearken to a time when our cities and towns worked better with the natural environment, in some ways. It is no wonder that these values would come into line, when new urbanism would meet naturalism and in many ways they would look the same. In some circles, the debate rages on about whether new urbanism is a form or if it is a function. These authors say the answer to this question is, decidedly, "yes."

# Master Planning: Applied New Urbanism

A concerted effort to maximize the openness of the neighborhood sidewalks, streets, paths, and common areas is intended to promote walking and community interaction, local economies, and ultimately, efficient transportation via a lessening of the need for cars. Civano is a place that has been inspirational for many of the people who have visited or purchased a home there. It is an attractive and comfortable place to walk outdoors through the mature desert landscape, bike to work, or spend a few hours in the dirt trails behind the community. One buzz-word that comes to mind when experiencing the first phase of Civano—homes with alleyways, few street-front garages, many front-porch areas, smooth sidewalks, gravel trails and human-scale streets that squeeze the automobile to bring business closer to home—new urbanism.

New urbanism attempts to heal the fractured connections of our landscape. This philosophy is bound to the different scales of geography and the fringe, the transition between places. New urbanism considers how the home connects to the school or the factory to the store. It is concerned about the poverty left in the wake of highways cut through once-congruent neighborhoods and the destruction of land bridges that allow for migration and genetic health of wild animals. New urbanism happens in our built areas, but its concerns stretch across the earth; for we take resources from the entire planet to support our urban areas. Below are some broadly-scoped items to consider when designing with nature, and human nature, in mind including details about how these items were addressed in Civano. Successful developers today and in the future will be able to make a good business case out of most, if not all of these items (and more) in the localities that they are considering to transform into human habitat. The others will fall to the way-side.

## CONSIDER DENSITY

As a point of reference, as of 2008, the LEED-Neighborhood Development pilot program requires a minimum density of seven units per acre and starts doling out points for higher density development. Civano, even with its often close-knit houses... is about five domiciles/acre. This figure is after removing the 30 percent minimum required open area that is mandated in the zoning rules for the area and approximately 50 acres for common park spaces, retention/recharge basins and 45 acres for the commercial center and easements.

Although the overall density of the project falls short of the LEED—Neighborhood Development pilot program goal, the community does have townhomes and "in law" suites that could be rented to a single person with few material needs and those units are not counted in the total. Apartments would be one excellent way to improve density numbers in a walking community where multi-family units were originally planned within the community, but apartments are still not being built at Civano. Single-family homes were the first to be built and occupied, and those stand-alone homeowners were the first voices to form the Civano Neighborhood Association's opinion.

When the first apartment was planned for development in the community, the homeowners shut the process down with vehement protest. The obvious aversion is the idea that apartments and other rental housing can bring down surrounding property values. However, the urban design of Civano yields a very comfortable environment where the homes themselves are generally arranged tightly together and the open spaces allow room to roam. The same would still be true even if there were a few apartment buildings in the mix. One of the challenges to scaling a project to allow for diversity is that many more kinds of housing must be made available. With more options come more opportunities for debate and friction. Early homeowner education may alleviate some of those concerns, but the bare fact remains that there is no way to please everyone at the same time—especially when considering the way our communities, and our homes are established and developed. Furthermore, project resources are always limited and can only extend so far to address concerns like consumer education. Thus, diversity is one of the most challenging aspects of design integration.

## DIVERSIFY USES

Biodiversity gives life options for specialization and flexibility for adaptation when the environment changes. Similarly, diversity in a community offers people many choices for work, recreation, socializing, and residence within a more geographically finite realm. Adding diversity of uses to community design increases the chances that individuals living there can stay in the community during their day-to-day activities, rather than drive elsewhere. Some of the intended results are decreased automotive travel, increased resident health and physical activity, theoretically more time for family and work, richer community involvement, and development of a more local economy.

A major lesson learned in Civano, which will be expressed on many levels throughout the text, is that the duality of design intent and construction phasing cannot be ignored. Civano was intended to have jobs on site, yet well over 1000 homes were built before the true commercial center even got its utilities or paved roads. This, in effect, renders Civano a community with very forward-thinking pedestrian-oriented design without anywhere local to go. Therefore, the place is not yet functioning as intended.

With the exception of a number of home-based businesses started there, the people who live in Civano now are, by necessity of location, auto commuters who work in the greater Tucson area. They sleep in Civano—a very beautiful master planned community which is currently functioning as a bedroom community to Tucson.

Be that as it may, the time will come when the commercial center is filled out and local home ownership can start to evolve to include people who live and work within the community. It will simply take a long time to happen. In retrospect, it would have been advantageous to the functioning of Civano for the commercial aspect to have been built in conjunction with the housing. Ideally, this would have been done in a ratio commensurate with the final design intentions. For example, given Civano's goal of one job on site for every other residence, that would mean that enough commercial property to support 50 jobs could have been developed for every 100 homes built and sold. If this suggestion were policy, Civano would have enough commercial property developed to support 500 jobs on site at the time of this writing.

In retrospect, it may have been too lofty a goal to achieve, ultimately assuming there will be approximately 1250 jobs in Civano. However, only time will tell if that is the case. What can be done is developers can plan ways to offer flexibility to stimulate community economic growth simultaneously with providing homes. One approach is to build one high-performance commercial building shell envelope to coincide with a number of homes built and sold. The shell can be built to LEED standards, or other reliable high-performance building standards, and offered as "build-to-suit" tenant improvement commercial sites.

If a superstore is accepted into the community that will require the whole shell, so be it. However, if a small boutique wants to buy into the project, that can be accommodated as well. Planners and project stakeholders can avoid strip-mall like development by scaling the size of building shells to match developmental goals and stylizing the façade with the architectural effects expected in the community. This approach splits the difference between going full-into developing a commercial site that may not prove profitable without eliminating the opportunity to allow a mixed-use, master-planned community to develop as originally intended.

## Residential and Commercial Diversity

There are many master-planned communities that offer a plethora of creature comforts such as golf courses, club houses, spas, and artificial lakes. The stand-alone single family residences or large multi-family buildings within these oft-gated communities may sometimes appear differentiated by their architectural details but are usually very similar in size, design, efficiency, accoutrements, and value. These communities are primarily comprised of partially-aligned demographics—people with like incomes and patterns of urban living. Although currently appealing to many, these homogenized communities do not offer a very good recipe for the kind of diversity that would support a local economy that draws greatly from a local labor pool, from the banker to the clerk.

One of the early goals defined for Civano was to provide different types of housing (including affordable housing) in different price ranges to house a diverse population. Ideally, a mixed-use, mixed-income urban neighborhood will be developed with a ratio of low, median, and high cost housing commensurate with the number of lower, median, and higher paying jobs that will be located on site. These ratios, in turn, are likely to reflect the ratio of lower, median, and higher income brackets that are represented in the region or the target market expected to be drawn to the project from non-local places. The implication of designing places to mix different people together has deeper sociological underpinnings, the likes of which these authors are not qualified to fully address. However, what one can be sure of is that attempting this type of design integration vastly adds to the complication of not only the physical aspects of the project itself, but also the marketing and community-building aspects as well. The design team should take this well into consideration when attempting mixed income and mixed use integration of their designs.

One of the major urban tragedies we are still recovering from today is the degradation of our American downtowns. As the automobile pushed our cities into sprawl patterns, lower rent strip-malls and small, disconnected commercial areas pulled businesses away from downtowns. Concurrently, outlaying bedroom communities worked as a kind of social filter that allowed people with more financial means to further disconnect and those who could not were left behind. The resulting decayed inner cities housed those with the least money, where there were no longer many jobs. Meanwhile, suburbanites unwittingly sentenced themselves to life behind the wheel commuting between home and scattered employment centers. Thus, to provide for a diverse population, the task of keeping everybody out of their vehicles and in the village requires the introduction of many different local job opportunities—like one may see in a vibrant downtown district.

## MIXED USE ZONING—COMING BACK HOME

**Moody:** Another aspect that had developed during the planning process was mixed-use and how much you could really blend uses. And I guess that was primarily my addition. I pushed hard for that. I personally am not a believer in segregated zoning because I think it depletes the enthusiasm and, you know, the kind of activity of a vital community. And it cuts down on face-to-face communication, et cetera, et cetera. And we wanted to do everything possible to create an environment in Civano where people would actually want to live and work at Civano. It wasn't ever going to be a requirement—you can't do that. But to create the kind of environment that would be conducive to people who wanted to live and work in the same place, raise their kids; you have the work become part of the family life. You know, it's just part of the community.

And so to be able to mix uses, even in small, fine-grained areas, to me was a very valuable thing to do. So it not only created the jobs to housing mixture, but it created the kind of architectural interest and diversity that I wanted to see in the community so that around the neighborhood center, for example, in the first neighborhood, I wanted to see like even three-story buildings or four-story buildings, you know, right around that ring. So have a lot of activity going on that would help support the kind of activity that might go on in the neighborhood center itself, become very alive, very vital, not just a place for people to live and then take off during the day and work.

Plus, just from a practical standpoint, you could reduce the amount of space given over to the automobile if you have mixed use, because at nighttime, residents could use their parking. Then in daytime, people could, you know, come in and use the same parking space, you wouldn't have double count. And I made a big deal out of that. And actually in the plan for the first neighborhood at least we were able to get the parking requirement cut in half and also farther away from people's, you know, place of residence and work.

Herein, one sees why integration must be carried out on as many levels, across every scale of the project that is possible. The idea that there would be jobs developed in Civano to limit vehicular travel led to the logical conclusion that there would be fewer vehicles in the community and thus fewer parking spaces needed overall. However, if one integrated aspect of the design is not concluded or is based on a faulty assumption, then the purpose of that integration may turn inward and become an integrated design challenge.

As stated, mixed-use zoning was incorporated into the Civano design with the goal of achieving one job for every other home built—2500 homes, 1250 jobs. Time will tell if those jobs actually come to fruition in Civano, but the planning for parking intends it to be so. But what if 1250 jobs are created, and only very few are occupied by Civano residents? Maybe the unanticipated number of auto commuters coming from outside the community will be forced to park in front of residents' homes. This may be balanced by the fact that Civano's residents, not working at the walking-distance jobs, would mostly be taking their cars out of the community to their jobs in the city. But Civano's streets were not designed for a double-load of vehicular travel as could be created during 'rush hour' given this scenario. Civano's streets are narrow and designed to squeeze the automobile to make room for the pedestrian. Now, only time will tell what the end result will be.

## TRANSPORTATION NETWORK

So much of what new urbanism seeks to restore is the human scale of our immediate communities and more natural linkages within our built environment. An easy way to visualize "human scale" compared to "car scale" would be to think about the distance a person can walk in 10 minutes compared to how far a car can travel in 10 minutes in an average-density urban setting. Assume a walking speed of 2.5 miles per hour, and a city driving speed average of 35 miles per hour for a car. In 10 minutes, a person will travel about 0.4 miles (just shy of a half city block) and the car will travel about 5.8 miles (about six city blocks). Incidentally, a car that gets 30 miles per gallon will emit approximately 4 pounds of carbon dioxide during that 10-minute jaunt and has several other direct and indirect costs associated with its operation-such as potential injury or death and environmental impacts.

Now ask what potential day-to-day resources (markets, schools, parks, employment, or mass-transportation centers, etc.) did the pedestrian pass in half of a city block in the average-density urban setting? Probably no more than the few stores constituting a strip-mall. Clearly, in the average western city a pedestrian is at a clear disadvantage and the automobile is almost a necessity. Of course, we still have to work with what we have. Designers of Civano and other walkable communities have thought about the human scale and how to maximize the resources available to pedestrians while still incorporating enough car infrastructure to connect to the larger region—due to our reliance on the existing car-scale environment surrounding the planned community. Listed below are a few of the techniques that were used at Civano and variants of strategies recognized by the LEED-Neighborhood Design pilot program and the Congress for the New Urbanism.

## Squeezing the automobile

It is amazing how many times people ask, "Is this a one-way street?" when first visiting Civano's phase I neighborhood. The roads are as narrow as the City of Tucson would allow and although two cars can easily cross paths in the opposite direction, most of the curbs have no vertical edge and are sloped to allow cars to move aside for larger vehicles.

The speed limit is a standard 25 mph, but in many areas of the neighborhood, drivers naturally go slower, feeling a bit squeezed by the narrow streets and desert-landscaped medians.

**Figure 3.5** Civano's rainwater harvesting landscape curbs are cut at an angle. This allows vehicles to maneuver through the narrow streets.
*Drawing courtesy of Justin Cupp.*

In a commuter world we theoretically need at least two parking spaces for each car. One for parking at home, and one for parking wherever else we may be parking at the time. The caution earlier about parking notwithstanding, mixed-use zoning allows for people to live very near where they work and shop, thus reducing the parking infrastructure needed for their automobile. Reduced street width and fewer parking spaces diminish the amount of asphalt required in the community, lessens the up-front costs of the street infrastructure, reduces the heat-island effect and lowers maintenance costs.

There were ideas offered early in the development of Civano about limiting in-community travel to only electric carts, bicycles, and other pedestrian-oriented travel. This would have required a place for people to park their cars on the periphery of the community. Similar to many idealistic designs presented early in the project, this one did not prove practical in application. Human scale is not only a consideration with regards to modes of transportation. However, squeezing the automobile, little by little, will help transition the built environment back to a more human-scale design. Civano and other new urbanist communities take one or two steps in that direction, but it will take time.

## Protect pedestrians

Most people have had the experience of walking along a sidewalk, like they are supposed to, and feeling in danger for their lives by the cars only a couple feet away on the street. It is almost as if the sidewalk is a buffer from vehicles for street side shops rather than a safe place for pedestrians. With narrower, slow roads, Civano already has a lot going for pedestrian safety. Additional features that protect pedestrians are desert landscaped buffers between the street and the sidewalk, sidewalks on both sides of most streets, bicycle paths and gravel paths that connect the community along routes away from any of the roads.

Remember the description of a 10-minute urban walk, half a city block. By creating more direct paths from residential areas to mixed use and commercial zones, pedestrians can often walk a shorter distance to their destinations than if they drove on the street—and in more natural-feeling surroundings. This increases both the density of potential resources in their human scale environment and the chances people will leave their cars at home.

## Include transit facilities

Mixed use zoning and commercial centers within walking distance of homes will keep people from leaving the community for many of their needs and wants, especially given today's predominant urban design. It is reasonable, however, to make it possible for residents of such communities to still leave their cars at home even when it is necessary to travel outside the neighborhood. Providing transit facilities such as bus and light rail stops as integrated features to the community continues the theme of supplying a diverse group of people diverse options to satisfy their needs without the use of cars.

## Support public transportation

Civano was built in a Tucson growth corridor, ahead of much of the growth. Houghton Road is the eastern most major road in the city grid street system that reaches from far north, all the way to Interstate 10 which borders the southern reaches of Tucson. A public transit option has not matured as of the time of this writing. There are bus stops within the community (Tucson has only a limited streetcar infrastructure located downtown) though no city busses stop in the Civano area as yet. School busses and assistive medical transports frequent the community, but it will require further development of the Houghton Corridor (see Chapter 1) before the area has enough population density to warrant mass public transportation.

The design of Civano as a walking community with no transit rekindles the statement that in order to create a performance or points-based green building program, a line must be drawn between what constitutes sprawl, and what is part of an established and ongoing regional development plan. Although Civano fits into the latter category, the time scale of development may not have allowed Civano to earn high marks in this regard with most up and coming green building programs. Civano bus stops are a very small portion of the urban design. The stops do not require much space, land, or infrastructure. This is a small price to pay to have the option for mass transit without the cost of a retrofit. It was most likely a good decision to not install larger, more expensive transit infrastructure given Tucson's lack of transit infrastructure. The investment would most likely have gone completely unused.

## CONNECT TO SURROUNDINGS

Ideally, our urban places would be well integrated with each other creating a public transportation infrastructure which would be least imposing on the human scale as possible, while still providing a practical means for people to move throughout the larger region. Mass transportation that has become ubiquitous in densely populated cities and megalopolises around the world are not yet practical in much of the American West. In the City of Tucson, there is a bus system, Sun Tran, and a small streetcar track that is being expanded along a single route through downtown. Sun Tran recorded nearly 600,000 more riders in 2007 than 2006, but as of yet, the plans to extend the route network through the Houghton corridor and Civano have not begun.

This leaves brave bicyclists and stoic motorists no other options for travel outside the community. As reviewed in Chapter 1, Civano is part of the Houghton Area Master Plan and no doubt will one day be surrounded by development and public transportation access, but until then Civanoites must either provide for themselves on site, or drive elsewhere to find much of their day-to-day needs.

## PUBLIC SPACES

Civano includes a wide variety of common spaces that typically embody the environmentally conscientious goals of the project. One of the prime strategies for increasing density and decreasing water use is designing homes with smaller yards. In several home designs, there is little more than a courtyard outside. Therefore, community gardens, solar heated pools, tennis courts, community centers, game fields, walking and biking paths, and playgrounds fill the gap. By consolidating recreational areas into common spaces, several environmental challenges can be addressed at once. One example is the reclaimed water system. It proved to be expensive and somewhat impractical to deliver reclaimed water to each residence in the first Civano neighborhood for watering small, desert yards. What has proven much more cost effective and technically practical is the use of city-reclaimed water on common landscaped areas that everyone shares.

## MAXIMIZE ACCESSIBILITY FOR ALL

Civano and other communities may be designed for walking, but pedestrians come in all forms. In a traditional urban setting, residents generally would be leaving their home and going to a public place, usually via car, and that public place should be equipped to service people of all levels of physical ability. In a pedestrian-oriented community, everything in between the home and the destination must be more carefully considered for accessibility. Consider wheelchairs, bicycles, skateboards, scooters, shoes, velomobiles, and the phenomenal gyroscopic two-wheeled Segway; there is an ever increasing number of pedestrian modes of transportation that will suit almost any walk of life. Thus, a community designed for pedestrian travel needs to be designed with consideration for accommodating as many of these modes as possible while still safely linking the residences and community resources.

## INVOLVE THE COMMUNITY

Most of us living in the west or other less densely populated areas are used to a suburban lifestyle. Bedroom communities are where we eat and sleep in-between leaving for work and recreation. Our culture has somewhat adapted to the environment which we live in and for so many of us, it is not a common experience to step out our front door, walk a few minutes, and find ourselves at a farmers market, community center, park or community sponsored coffee social.

Culture, like a neighborhood, needs to be built and cared for. A pedestrian-oriented urban design is only part of the equation. There are many other aspects of integration that are not on the forefront of design, such as Covenants Conditions and Restrictions (CC&Rs). CC&Rs are rules and limitations put on a specific group of housing units, homes and/or community common areas. CC&Rs help shape a community in more ways than predicating which colors homes can be painted or how many cars can be parked at the curb. The method in which CC&Rs are enforced, the type of interactions that are stimulated between neighbors, and whether the rules are logical for the place all can affect a neighborhood culture.

Unfortunately, most CC&Rs are not more than a common boilerplate legal document that some developers hand out to their newly built communities as if one size fits all. As we specialize our communities, make them mixed use and invite in all walks of life, the common CC&Rs simply may not fit. Civano's first neighborhood is a more colorful place because its CC&Rs offer a more varied palate of color choices than many neighborhoods. This one small bit of leniency has brought much character to

Civano. Other aspects of the community rules and means of enforcement have been evolving to be more appropriate for the specific community of Civano as different people participate in the homeowners association activities.

**Rollins:** Another missing part of Civano is the [acknowledgement] that it is the nation's first, and I think still the only, sustainable new urbanist development. There are lots of new urbanist developments like Seaside and Kentlands, the re-creation of pre-World War II neighborhoods that we all grew up in. And so Civano had that element, I would say more of a modified new urbanist development. But Civano also had as its mandate its original purpose which was sustainability, or the use of resources to the degree where you are meeting today's needs without sacrificing the ability of future generations to meet their needs. And that's paraphrasing the United Nations' explanation of sustainability. So putting those two together was just a daunting task, and it had never been done before.

But primarily, sustainability was the mandate, as well as its public/private partnership with the City of Tucson and the state and many, many different bureaus in the state. And all coming under the control of the city. And to accomplish that, the city signed a development agreement with the developers and established guidelines and goals to meet on the level of energy use, resource management, water conservation, solid waste management, things of this nature. And they put it all into a program which they called the IMPACT standards, which they would then monitor to make sure that the guidelines were kept.

And so the city has always been an integral part of Civano and has—it has been a mixed blessing, to put it mildly. It's been difficult for the developers in many ways, but it's also guaranteed that the original goals of Civano were maintained. Anyway, that was the original name of it, The Tucson Solar Village. And it was only in later years when the city and others got involved with Civano that they changed the name to Civano because the goals had broadened considerably beyond just the use of solar power. So to call it solar village became somewhat limiting. So they changed the name to Civano. And "Civano" really comes from the Hohokam civilization. The Civano period of the Hohokam civilization was a period when supposedly the indigenous people lived lightly on the land—the golden era of the Hohokam civilization: The Civano period.

So because of these goals, Civano attracts a certain number of people who hold those values. They want to come here because they hold the values of living more lightly on the land, if you will. And over the years, Civano has more than met the goals of sustainability that were set in 1996 with the first development agreement. And in fact, other developments have been hugely impacted by what Civano did, including the new Rio Nuevo that's going to be built in downtown Tucson and the Armory Park development very much improved on and carried forward with the Civano impact statements.

So it has had its impact. And while it has struggled financially, it has achieved in many ways its goal of influencing the way development was done and is done in the southern Arizona area.

At every decision, a project design team is faced with the constraints of what already exists versus the resources available to assert change. Existing infrastructure, best practices, developed technologies, knowledge, and conceptions all play a part in decisions about innovation that must be made during the course of a project. Ideally, a balance is achieved that can be fruitful, replicable, and advanced whenever possible. On one side of the scale is ideology, on the other is reality of a budget. If too much is placed on the ideology side, the balance will tip and money will come up short. If up-front cost savings is overweighed, good ideas and real cultural progress toward creating better practices, technologies, infrastructure and knowledge may be forfeited at some unknown future expense.

**Cook:** There were several opportunities created by the team that managed this process. The project director, Will Orr was working out of the city manager's office and The Metropolitan Energy Commission was overseeing it. They had an actual energy office in the city. The project team created opportunities for several general meetings. And there were some charrettes that involved the public.

And so we were able to provide input into this design process, this planned development, and this potential community. And our little group, I had an influence in really trying to push the envelope here and make this a significant development. We went beyond just, you know, the buildings. We did look at the transportation issues, we looked at the employment issues. We even talked about the possibility of creating a local currency. So we were looking even into the deep economics of how you would create this—this sustainable community. This piece of property was owned by the state lands department. And they owned a lot of property along the Houghton corridor. And their idea was to create an urban lands development model that could be applied to a lot of their land along that corridor. That was the original vision.

Of course that didn't really play out. We didn't really see them take that aggressive role in trying to support the wider sustainability goals except for the basic water energy and goals. And those were the two that probably had the greatest impact in terms of raising building standards—much higher standards than were required in the City of Tucson. And since this was in the City of Tucson, just on the edge, the city had an interest in it and we were able to craft the Sustainable Energy Standard, which were significantly improved modifications of the energy code that would be applied to any structures built in this development. And it was enforced through the permitting process. So the Civano Sustainable Energy Standard became one of the most effective parts of what we were able to accomplish. Because one thing that should be remembered about Civano is that this development was originally called the Tucson Solar Village Project. So the idea was to apply solar energy to the community.

After phase one of the planning process, we came up with some very specific areas where we would require performance targets: energy, water, solid waste, air, employment. And just look at how aggressive the employment target was. "One job will be provided for every two residential units built." That was a goal that never was even closely achieved. I don't think they really understood the dynamics of how you create a village economy, a critical mass of businesses and development. I don't know—do—are you living up there right now?

**McDonald:** I wish.

**Cook:** Oh, yeah. But you go out there to work because that's where, yep, Al's office is.

**McDonald:** Or cry in my beer.

Indeed, not being able to see Civano is to cry in one's beer. For some, seeing Civano is also reason to cry, for they knew all too well the full extent of what it was first visualized to be. Bob Cook's perspective is a clear expression of the balanced thought that in spite of Civano's shortcomings, it was a leap forward. Why and how the dynamic and pure idea of a much more sustainable village became watered down is best explored by the different people involved with Civano's development and the diverse perspectives involved.

From Paul Rollins's interview with Wayne Moody:

**Moody:** A very interesting story, which I haven't shared with too many people, is about the landscaping. Coming from kind of an environmentalist background and an earth-centered background, I was very concerned about the potential destruction to the landscape on Civano. And even though we did really good environmental work and only selected land to develop that was not very well suited to vegetation, there were still a lot of trees, a lot of saguaro, and so forth that had to be moved or destroyed. And so actually as it came close to time to start grading for Civano, I got really nervous. It was very difficult for me, almost to the point of tears, because it even kind of gets me when I talk about it now.

So I had a close friend, close personal friend who was an intuitive, psychic channel-type person. And so she and I came out to the site one evening, late, there was nobody else around. We had a conversation with the plants. I was not sufficiently developed myself to do that, so I used a surrogate to translate for me. And very interesting information came through on that. There was the big diva of this whole area that lives in the Catalinas that showed up, huge being, and said, "I know what you're trying to accomplish; and it is good. And if you will make it possible to harvest the trees and so fauna that are here, we will thrive better than you can ever imagine."

And so that made me feel good. So then I got together with Les Shipley; I said Les, we got to, you know, whatever possible to preserve these trees. Well, Les, bless his heart, bless his soul, developed all new techniques for harvesting trees.

**Figure 3.6** Native trees were boxed and reused in neighborhood landscaping with a 90 percent success rate.

I think there were 1400 trees that were harvested and saved in the first neighborhood alone. And he put them in boxes (see Figure 3.6). It's beautiful.

And so there's a great photograph that I love that shows all these trees in boxes, right? You look at Civano today, and it is totally lush; it's three times as lush in the neighborhood as it was before the neighborhood was here. So I feel good about that and I've used that practice in every project since where you talk to nature.

Wayne Moody passed away not many years after his work in planning Civano. Before he died, he finished the Tucson co-housing development of Milagro—a 28-home site with mostly open space. Cars do not enter near the home areas but are parked in an adjacent community lot. The homes are made in part from the earth on site. Natural wetlands treat wastewater and irrigate the landscaping. It is a beautiful, natural setting and it harbors a community that is small, yet connected and seemingly more alike in purpose. The place feels close to Earth.

Wayne Moody's spiritual beliefs played a role in his land development choices. He brought methodologies to the table that were not mainstream and mixed them with cutting edge development practices now called "neo traditional" or "new urbanism." Whereas Milagro is of a small scale where it was statistically more probable to find enough purchasers who identified with Wayne Moody and his personalized approach, Civano's huge scale required many more personalities and philosophical compromise on the path to being built. However, of the different planning approaches there is one in particular that was not used in Civano, yet is now viewed as an indispensable tool for master planned community design integration and execution—the plan book.

**Figure 3.7** The Nichols' Tracking Solar Oven, located in the Milagro co-houseing development, is dedicated to Wayne Moody.

## THE PLAN BOOK THAT NEVER WAS

**Rayburn:** We never developed a true plan book and I think this is why the project went awry after Fannie Mae took over day-to-day management of Civano, and none to the original development team remained.

The Civano planning process was advanced in terms of integration and methodology, but only condensed conceptual design ideas derived from the collaborative effort as far as the charrette level. This practice yielded a charrette book that held large-view concepts and spatial relationship guidelines on the scale of something like a zoning map. Taken a step further, the concepts of the charrette book should then be further developed to look at design concepts for the project with finer detail. This yields a plan book.

Once completed, the plan book (or design book) will present itself as a whole series of pictures, schematics, and matrices that give the builders a realm of choices and help them understand the relationships, consequences, and constraints of each design choice they consider. It is like drawing boundaries and shapes but leaving the space in between somewhat fluid. The plan book describes key spatial relationships in an attempt to create the least design limitations possible to achieve a desired hierarchy of scales, and overall "look and feel" that is generally consistent, but allows for great variation in details. For example, it may describe setbacks and building heights along a boulevard that would complement the scale of the road. Thus, one would experience a different set of physical relationships depending on which kind of street he was on.

The plan book also may be used to define parameters for appropriate detailing, such as whether or not one can have a porch on a dwelling built in a specific area. The ultimate goal is to create a design vernacular on different scales throughout the development. To that end, the design book also helps define the hierarchy of public and open space. This allows designers to understand when there is the necessity for a square or green space, where these spaces are located, and in which natures they should connect the neighborhood together.

**Rayburn:** This lack of a codified set of design principles compromised the project in terms of design control. Typically in new urbanist design you have a master planner who helps guide the creation of a complete set of design guidelines, which is just like a matrix that you can give to the builder and say, "There are lots of choices inside this box of design guidelines. You, the builder, can make choices, but just stay inside this box of guidelines."

Ironically, there was a book developed in 1999 by the Tucson Institute for Sustainable Communities called: *Sustainable Design: A Planbook for Sonoran Desert Dwellings*. This plan book was developed with the intention of exploring how Civano's homes achieved resource conservation goals as a springboard for exploring the concept of sustainable design as an alternate to conventional design processes. This type of reverse engineered plan book goes to show that Civano was ahead of its time in results, if not methodology.

## NEW URBANISM'S CODE

Form based planning is new urbanism's answer to building code language and an extension of what goes into a plan book. Chapter 1 had a brief introduction to some of the ideas presented in the Charter for the New Urbanism wherein the charter was described as a form of pattern language from which one can extrapolate design generalities into specific meaning given a specific context. As the charrette book comes before the plan book, so the charter comes before the charrette. But why is it that, so far, there is really no specific code language for new urbanism? A quick look at building code language compared to pattern, or form-based language will help make a distinction.

A section of Chapter 4, Section 402 of the 2006 International Energy Conservation Code (IECC) reads as follows:

> **402.1.2 R-value computation.** Insulation material used in layers, such as framing cavity insulation and insulating sheathing, shall be summed to compute the component R-value. The manufacturer's settled R-value shall be used for blown insulation. Computed R-values shall not include an R-value for other building materials or air films.
>
> **402.1.3 U-factor alternative.** An assembly with a U-factor equal to or less than that specified in Table 402.1.3 shall be … etc.

Notice that the code language is finite. It is something that can be checked using rulers, numbers-empirical evidence. Comparatively, here is a section from the Charter, Chapter 12 which speaks about creating walkable communities:

> Traffic planning techniques of "assigning" traffic—or assigning the projected quantity of travel on specific routes, based on notions such as how many trips a typical family might make each day—reveal the important advantages of a highly connected network:
>
> ■ Local traffic, which comprises 70 percent of all vehicular traffic, stays local. With the connected street network, local traffic uses small local streets and never enters the major arterial system. By contrast, the conventional suburban pattern of cul-de-sacs

feeding into a main arterial compels all drivers into the arterial system. This focusing of all traffic onto arterial highways produces intersection congestion even in low density developments ... etc.

The difference between the form based or pattern language and the building code language is fairly clear. While building code language speaks mostly in absolutes and specifics, the pattern language sets a framework in generality and based in assumptions about how people will behave. Logic follows that the specificity of code language makes for a more predictable environment for builders and designers. They know that if they follow the code, the will have to use specific types of materials, specific techniques and configurations—all of which will be inspected for accuracy. More to the point, it requires a huge spectrum of documents to adequately describe how a development project should fit and interact as a component of its region, down to which kind of lockset will be put on the front door of a specific home at a specific address.

For the master planning process to work in a new urbanism setting, the vision of developers must extend beyond just the complexities of the lot they intend to build and the immediate connections to utilities and roads. It requires them to question what kind of boundaries and form the development should take for itself. Their answers would show careful reflection of how the development would affect and be effected by what is already around it. This, in turn, will determine how its individual components of the development and surroundings will relate to each other. Then, these relationships will determine what mix of uses in the development would create the least overall impact and maximum health and safety to reach the highest potential for the community of people who will one day occupy the homes and businesses therein.

Sometimes, even the best design ideas cannot feasibly integrate into the existing paradigm, which in turn, will effect considerations of project scale in terms of size, complexity, and innovation. However, once the mix of uses for the development is determined, the master plan is set and the plan book has been developed, details for consideration become noticeably more complex with detailed architectural and mechanical concerns—building code concerns. Somewhere in between pattern language and building code is the fuzzy area where pattern language and code language meet. Ultimately, urban planning tends to take a look from the large scale and incrementally reduce it down into system relationships. Meanwhile, code language describes small- scale bits and pieces fit together to create the systems. The planning process defines the design parameters and conditions that will result in the embodiment of the project's ideological goals, most likely expressed most purely in the early charette processes.

# 4

# IMPACTS AND ADJUSTMENTS

## The Basics of High Performance, a.k.a. Green

*Extraordinary claims require extraordinary evidence.*

—CARL SAGAN

Scientifically, the color green denotes a color of visible light measured at a wavelength range of approximately 520–565 nm. Because people have measured the wavelengths of light, when one observes a green light changing to an indigo color, we know that means the wavelength has shortened by about 100 nm. But what if we didn't measure? The fact that the light changed to indigo would mean nothing quantifiable and green would merely be an idea, just a nice, non-indigo color. The same is true for the green building.

Green is the symbol of the movement, a way of living, and the color of our environmental aspirations. As such, green is a beautiful thing. But without measurement, the greenness of a product or service is merely a hunch, completely void of scientific merit and impossible to manage effectively. Technically speaking, to call such a thing green is only metaphorical. As a word used to describe the environmental performance of a building, vehicle, paint, process, or any product without a hue reflecting or emanating light at approximately 520–565 nm, green is inaccurate and inappropriate.

Performance denotes a measurement compared to a known value. The quantities that determine environmental performance have to do with measurable inputs and outputs. MPG, Btu's, Watts, board feet, tons, gallons, kilowatt hours, therms, cubic yards, squares, millimeters, parts per million, rad, nanometers—all terms of measurement that may apply in some way to understanding environmental performance. Performance, in the purest sense, is a neutral concept. In order to understand if the performance of a system is good or ba there must be a target or goal that is defined in the same terms of measurement that denote performance.

The planners of Civano realized this and mandated a process for measuring, monitoring, and evolving the standards guiding the development of Civano's performance. This one feature of Civano, a mandate to monitor the performance of the project until substantial completion is achieved, closes the loop on sustainable project management and makes the Civano project stand alone as a source of data about high performance construction and community development.

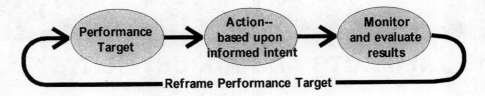

**Figure 4.1** Impact process flow chart.

When measuring performance in terms of energy used by a machine or the sum energy of a process, it is necessary to have a point of reference, a baseline measurement from which changes in performance can be measured. The following discussion with Al Nichols tells the story of necessary efforts to define the baseline to which Civano buildings' energy consumption performance would be compared.

**McDonald:** You said that during your time in one of the Tucson Solar Village commission meetings, you said it was all about making a standard. Do you mean a building standard?

**Nichols:** Well, they had a standard. They had come up with a standard that said that they needed to save 65 percent on energy. That was based on the U of A study by Dr. Nader V. Chalfoun, University of Arizona CAPLA, School of Architecture.

**McDonald:** 65 percent of what?

**Nichols:** Ah, 65 percent of the energy used by 1960s homes. They had evaluated them; run math models; and said, "You know, by doing it right, we could drop the energy by 65 percent." And that was heating/cooling energy. It got down to a point where we had had all kinds of ideas. I mean crazy ideas were floating around. And so I threw a couple of my own out. I said let's give everybody a 20-amp breaker, a 3/8 gas line, and let them build whatever they want. And of course that was silly. Then they wanted all the homes to be properly oriented, and that's pretty impossible if you're going to do any kind of mass-production building. And it was obvious that we couldn't do anything but mass-production building because making it replicable is part of the project. We're going to build 2500 homes.

**McDonald:** You know, a lot of people talk about orienting houses and whatnot. As a professional engineer, what do you really think about that?

**Nichols:** Well, that's the study. You see, the study, back in the—in the '90s was all surrounding the ranch-type home, which was prevalent in the '60s. Those are essentially long, skinny houses and single story. One of those improperly oriented would make a difference of up to 30 percent in the energy use. And that's true. But when we start looking at the modern home, and particularly model homes, they are practically square. You don't see homes like we used to build. So …

**McDonald:** You mean today's architecture?

**Nichols:** The style, sure, but more the shape. They were long, rambling, low-roof homes. Now they are square, high-ceiling homes; you know, ceilings now 10 feet minimum. In a ranch home, they were 8 feet. So, the whole housing style had changed since the study was done.

At one point, we commissioned Katharine Kent, P.E. of Kent Engineering, now owner of Tucson's Solar Store, to do a study using CalPass.

**McDonald:** What is that?

**Nichols:** CalPass is simulation program developed, I guess, mostly for California because of its Title 24. It's a simulation program to show compliance and energy use. So she took an upscale home, that is upscale insulation, good glass, good walls, good roof, took them square, 1:2 ratio, 1:5 ratio, 2:1 ratio, and that spec and then rotated them in all directions and counted up the energy use. And in all cases in all different orientations, there was only about a 5 percent difference. Because a better-built home, better glass isn't as orientation specific. And there is a report on that. She completed 500 simulations.

And what that tells us is that with a 5 percent variance, we can allow the builder to build an upgraded home where they want to build and how they want to build them and build the same home on any orientation on any lot and the penalty is small.

**McDonald:** So the secured building envelope really does make up the difference?

**Nichols:** It really makes that difference. And it's a hard concept. I mean, we're building with a lot of hard concepts here. And one of them has been forever, this orientation concept. So through Katharine's study, it was clear that that was true at the time but isn't true in our planning to do this.

So again, I don't remember the chicken and the egg here, but at one other point, we as a company were commissioned through a grant through TEP to put together a team outside the Metropolitan Energy Commission to evaluate. So, it became obvious during some of our subcommittee meetings that we now had a baseline. That is to say, about a year and a half or two years before joining the commission, through the code committee, through the subcommittee of code committee, which I chaired, we had come up with a minimum code for all of Pima County and Tucson, based on the model energy code, which is a national code. And it was not a great upgrade. I mean, it wasn't terrific in those days. It was pretty soft. But it had established a baseline. That is, we now had a measuring stick. With the original concept of 65 percent below the '60s homes, there was no measuring stick. There was no way to enforce, there was no way to measure, no way to give the builder a method of compliance. Okay. The builders have to have a method of compliance so you can go forward. So I said, "You know what, how about we take the model energy code and just amend it?" Which is what we ended up doing.

The other thing that Katharine had done was a test through CalPass which helped us determine that the model energy code, even though it wasn't very stringent, at least improved energy efficiency by about 15 percent. The day we passed the model energy code, the builders started going double-pane glass, clear, and 2-by-6 walls instead of 2-by-4 walls. And so just because they— they knew they were going to be tested on this model energy code, they just began automatically upgrading the homes—and at very low cost. And it was a big fight even to get that. But we got it, okay, because we passed the model energy code.

So we said if we take the 65 percent factor, and that the model energy code is improved by 15 percent building, we're going to change Civano's Energy Standard goal to 50 percent above (more efficient) model energy code. So again, using CalPass, we tested heating-degree days, because heating-degree days is all that the model energy code recognized. Because model energy code is—it's interesting. The model energy code is not orientation-specific.

The general steps to effectively manage performance are few: determine a baseline in measurable terms, create a goal measurable in the same terms as the baseline, monitor progress, and adjust strategies as necessary. These steps sound simple enough, but getting there can be complicated. For Civano, defining the baseline took creativity and extensive simulation testing to describe in quantifiable terms, but once that was determined, the goal of improving Civano home performance by 50 percent from the baseline value was a tangible performance goal. The tools that were offered to designers and builders to determine design compliance are free and relatively simple Department of Energy software. But what about monitoring the actual performance of the project?

# Auditing the Project, Energy and Water

The Integrated Method of Performance and Cost Tracking System—IMPACT System is a means of organizing resource efficiency goals and stakeholder cooperation for sustainable community development and for measuring progress toward those goals over time. It is intended to be a cyclical process that:

- Is grounded on metropolitan Tucson baseline conditions that are normally documented and periodically updated by community organizations.
- Is responsive to community policy priorities that will change over time.
- Uses performance targets and specific requirements that exceed baseline conditions without detrimental cost penalties.
- Uses collaboration among stakeholders to reach common goals.
- Measures development performance and costs to evaluate target achievement.

- Enables revisions as baseline conditions improve, and as new targets become technically and economically feasible *(Revised Civano Memorandum of Understanding Appendix A)*.

Each year, Al Nichols Engineering, Inc. (ANE, Inc.) conducts an energy and water audit of Civano to comply with part of the Civano MOU's IMPACT process. Surveys are sent to 200 households in both the greater Tucson region and Civano. Survey respondents supply ANE, Inc. with physical information about their homes such as year built, square footage, energy sources (gas, electric, PV, etc.), and a signed release form that allows access to the household utility bills for a 10-year period for the purpose of this study.

Each year, the baseline is monitored by retrieving energy and water use data from Tucson-at-large homes to which Civano homes are compared. As estimated, those energy code homes built after 1996 tend to be approximately 15 percent more efficient to heat and cool than those built before energy codes and Civano is performing extremely well, at nearly 50 percent increased performance over the average baseline home. The following several sections outline energy use in Civano and the City of Tucson as a method that could be used in other performance-driven developments.

# Characteristics of the 2007 Energy and Water Use Monitoring Study

The most recent 12-month study period to the time of this writing was from January 2007 to December 2007. Housing data was collected and matched with all utilities for each given home. The study discounts annual weather functions that might otherwise alter the comparative results. Data for Civano and Tucson-at-large homes was collected from Tucson Electric Power Company, Southwest Gas Company, and Tucson Water based on voluntary participation in the study by Civano and Tucson homeowners. Participation in Civano's energy audit is always on a voluntary basis and utility companies are very careful to protect homeowner privacy. Previous releases plus new releases from respondents who replied to postings around Civano resulted in 70 participants contributing to the current study; average home size was 1927 square feet.

Surveys were not sent to homes known to have pools or spas because these luxuries are very water and energy consumptive (if heated with electricity or gas). An average size pool can have a gas heater equal or greater to the Btu rating of the furnace that heats an average sized home. It is sometimes challenging to separate energy and water used in a pool or spa from the energy and water used in a home or landscaping without specific metering of the pool faucet and heater. Therefore, excluding these homes from the study increased the chances that the building and landscape are being accurately targeted in the audit, not the pools.

Homes with an uncommon amount of solar photovoltaic energy and net zero homes are also not included because energy offset by solar power does not signify an efficient building as defined in the building codes. Net zero, (or grid-neutral) homes cancel out their energy demand with solar energy. Therefore, homes with much more solar energy than the average home in Civano could produce "false efficiency" levels, depending on how they are metered.

Homes for the Tucson sample typically include those built before the first energy codes were enacted (pre-1996 city homes), and homes built two years after the first energy codes were enacted (1998 to 2005 city homes). Names and addresses were obtained from Pima County tax records. A program then was used to randomly choose homes built during target years and release/information forms were mailed to 200 residents. A number of release forms from earlier years were still valid. When combined with current year respondents, the cumulative non-Civano sample size was 48 relevant responses. Of these, water data from Tucson Water and energy data returned from Tucson Electric Power and Southwest Gas company provided samples for 48 homes with 12 months of data; 32 homes returning data were constructed prior to 1996 (pre baseline energy code) and 16 were constructed during 1998-2004 under the IECC. The reported average Tucson home square footage from this group is 1726 square feet, ranging between 1028 and 2900 square feet.

Energy bills from each cohort were examined by month, and energy use evaluated and reported in source kBtu/sf/yr (1000 British thermal units per square foot, per year). Source energy is the energy that is actually produced at the power generation facility. Since energy is used for electrical transmission and transformation into the 120 volt alternating current (AC) used in most homes, measurements only taken at the home would not constitute the actual energy produced for that home.

## CORRELATIONS

Source energy produced from Tucson Electric Power is approximately 3.1 times the energy read at the home's kilowatt meter. Natural gas requires much less energy to transport in comparison to what gets used at the home, making its multiplier only 1.11 of what is read at the home's gas meter. Due to the parameters of Tucson Electric Power's energy production facilities, approximately 2.2 pounds of $CO_2$ are released per kWh of electrical energy in Tucson that requires the combustion of approximately 1 pound of coal and one-time use of approximately 0.5 gallons of water. In other terms, just more than 67 pounds of CO2 are released per therm of coal-powered electrical energy. The national average is approximately 11 pounds of $CO_2$ released per therm of natural gas. In some ways, gas is a less environmentally impactful way to heat a home and water (compared to coal-fired electricity production).

The envelope of a building is the sum of the parts that are physically in between outside, and inside—the exterior walls, windows, doors, roof, and floors. These are the parts which physically transfer thermal energy in or out of the building at a rate which is inversely proportional to the quality and amount of insulating material within the envelope. The energy component of the SES targets the building envelope insulation and the efficiency of the mechanical systems. Therefore, it is necessary to determine how much energy is used to heat and cool the residences, and separate that value from other energy consumption. The energy required for heating and cooling the homes was determined by averaging the base loads (or plug loads) for each month. The calculated base loads were then eliminated to reveal the heating or cooling energy for the month. Base loads come from those devices such as televisions, refrigerators, and computers that use approximately the same amount of energy throughout the year.

Base loads in ANE, Inc's audit of Civano are calculated using Tucson Electric Power Company's method for base-load calculation: the lowest monthly energy use found during March or April is averaged together with the lowest of the two months of October or November. The resulting average number is utilized as the base calculation for the sample. As an evaluation measure, this procedure assumes little or no heating and cooling is required in this region for the months of March, April and October, November. Mostly likely, there is probably a small amount of both heating and cooling take place during these mild months of transition from summer to winter and vice versa. This baseline procedure will almost always produce at least one month with negative numbers. In the latter case, some of the energy attributed to base load would therefore actually be heating and/or cooling energy.

Some base loads, such as incandescent lighting, will put heat into a home and create perturbations in the heating load or cooling load of a building. Thus, some electrically heated and cooled homes will shift some of the heating energy used in winter into the lighting component of the base load. In other words, the heat from lighting takes pressure off of the home heating system. Theoretically, this effect would be reversed in the summer when the heat from the lighting base load component creates a larger air conditioning requirement. These kinds of unknown quantities (without a full, expensive audit) are part of what is difficult to discern simply using the utility data base load calculation.

For the purpose of the Civano audit, dual fuel homes are homes that use both electricity and natural gas. Typically, a home with natural gas uses the gas for space heating, water heating, and occasionally clothes drying and/or cooking—essentially any process that yields heat as its primary function. All-electric homes use electricity for all of these functions. Some of the homes in the study have dual fuel status, but the gas is only used for the small load of cooking. When reviewing utility bills, these homes are typically easy to identify, having gas bills much too small to account for space or water heating from gas.

Given the above reasoning, base use is difficult to measure with exactness without using meters on the heating, ventilation and air conditioning (HVAC) equipment, and subtracting their values from total consumption; but the method followed here results in a good enough approximation to determine if there are major shortcomings to the Sustainable Energy Standard implementation or process. The results of the Civano utility data audit consistently show that there is much less difference in base loads between the baseline homes and Civano homes than there is in the heating and cooling energy consumption between the two cohorts. This suggests that people generally enjoy the same electronic appliances and gadgetry regardless of their home's construction type, but the envelope performance varies greatly.

## EVALUATION OF 2007 ENERGY USE

To measure improvement in energy use between the different sample sections of this study, Civano phase I homes, Civano phase II homes and 1998-2004 Tucson homes were compared to the baseline of pre-1996 Tucson homes. The reason for separating post 1998 homes from pre-1996 homes is that homes in the Tucson region became subject to energy codes for the first time in 1996. It is assumed that it took approximately two years for the effects of this energy code to come fully into effect. Results for total energy use for cooling and heating energy for 2006–2007 are given in Figure 4.2.

The results for 2007 are that homes in the study built under the International Energy Conservation Code (IECC) use 25 percent less energy for heating and cooling and 8 percent less energy total when compared to homes built before the energy code was in force. Civano phase I homes (Civano I), built to the Sustainable Energy Standard, perform better, reducing heating and cooling consumption by 49 percent and overall energy consumption by 39 percent over homes built before the inception of the IECC in Tucson. The most recent Civano homes, Phase II being built by Pulte Homes, are showing increased performance in heating and cooling over 'baseline' homes at 40 percent savings with an overall 18 percent improvement in energy consumption when compared to the baseline Tucson home.

Although phase II homes are approximately 10 percent shy of the Sustainable Energy Standard (SES) energy goals, the homes are operating at a substantially increased efficiency and a small percentage of the overall phase II homes had been occupied for a year by the 2007 study. Until there is a larger sample (as expected in the 2008 study) it is difficult to ascertain if this difference is due to behavior of inhabitants or home design. The overall success of Civano is determined in aggregate, but if it is found that Pulte's homes are not meeting the standard, they will be notified, strategies to bringing their designs to compliance will be developed, and adjustments will be made.

Figure 4.3 gives a representation of total heating and cooling energy consumption for all Civano homes in the sample, IECC homes and pre-energy code homes. Civano homes (phases I and II) used an average of 28 percent of their energy on heating and cooling and 72 percent on base loads. IECC homes in our sample also use an average of 30 percent of their energy on heating and cooling and 70 percent on base loads. However, pre-96 (thus pre-IECC) homes use an average of 36 percent of their energy on heating and cooling and 64 percent on base loads.

| Overall and Heating/Cooling Energy Use Reductions | | | | |
|---|---|---|---|---|
| Sample | Total use (kBtu/sf/yr) | heating/cooling energy (kBtu/sf/yr) | Improvement Total Performance | Improvement Heating/Cooling |
| Civano I | 63.8 | 19.6 | 39% | 49% |
| Civano II | 87.45 | 22.76 | 17% | 40% |
| Tucson 98-99 Homes | 96.6 | 28.7 | 8% | 25% |
| Tucson Pre-1996 Homes | 105 | 38.2 | BASELINE | |

**Figure 4.2** Total energy use for cooling and heating energy for 2006–2007.

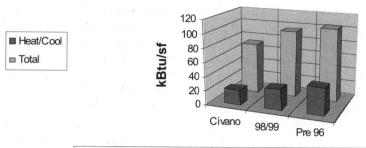

|  | Civano | 98/99 | Pre 96 |
|---|---|---|---|
| ■ Heat/Cool | 20.86 | 28.72 | 38.17 |
| ▢ Total | 73.28 | 96.64 | 105.09 |

**Figure 4.3** Energy use by type.

The graph in Figure 4.4 shows total energy as average energy use in kBtu/square foot/month. The graph shows two energy use peaks arising from seasonal energy use for heating and cooling. For the purposes of this study, dual fuel and all-electric homes are considered in aggregate. So, where the whole summer cooling load peak is comprised of electricity use, the winter heating load peak is comprised of electricity or natural gas use.

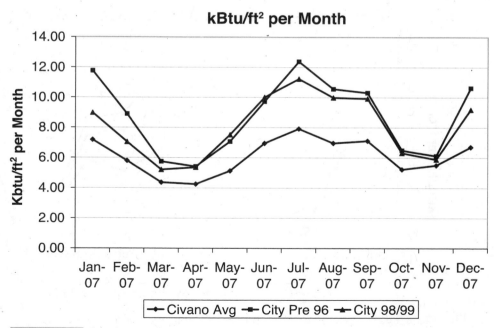

**Figure 4.4** Average energy use per month.

Figure 4.5 illustrates that the daily habits of Civano residents (relating to base loads) are very similar to those of the general population. Civano homes' consistently lower use of kBtu/ft$^2$ is due to increased efficiency of the building envelope and the use of solar energy. Behavioral differences may also play a role, though that cannot be corroborated within the parameters of this study. Figure 4.6 is a graph which illustrates this wide difference in behaviors. Even within one cohort, Civano homes' energy use varies from less than 25 kbtu/ft$^2$ to over 126 kbtu/ft$^2$. This 4:1 ratio is a consistent observation, suggesting that behavioral differences and number of people living in the home are significant factors in variable energy use among Tucson households.

It would be potentially very enlightening to conduct a full survey and audit that would ask participants to give more information about their homes and family to offer insight about factors behind consumption in our country. Perhaps there are key behaviors or factors that lead to high-end consumption rates that could be alleviated without interfering with individuals' freedom—such as lighting occupancy sensors or programmable thermostats that shut the HVAC systems off when the occupants are not home. Furthermore, the carbon footprint of a person living with other people tends to be smaller than that of a person living alone. Knowing how many occupants are living in the subject audit home would be most useful to deepening the context of energy and water performance, but this is not as simple as it may seem.

ANE, Inc. is released to draw 10 years worth of data given one release form, so long as the same person who signed the form lives at the residence in question. Early studies asked for more information, but as the years go by, families evolve and render

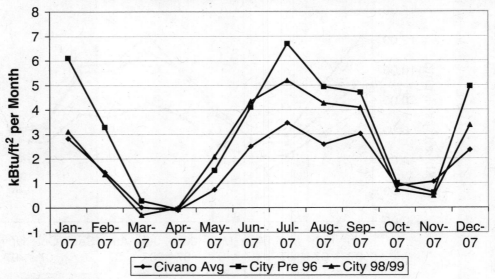

**Figure 4.5** Daily energy use habits of Civano residents.

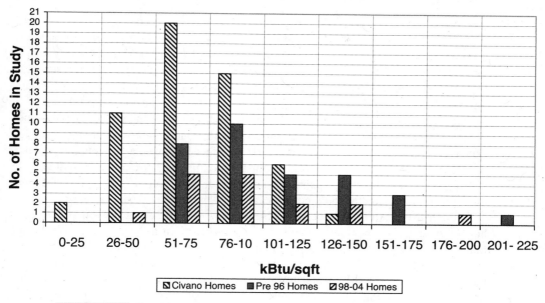

**Figure 4.6** Total energy use comparison.

some demographic data inaccurate. Thus, later year audits focus almost completely on utility data, with the understanding that there will be approximately a 4:1 difference in energy performance between similar design homes. It is currently impossible to do a blind study due to utility privacy protection practices. A blind study would benefit this type of research because there may be a correlation between the energy and water use behavior of people who would voluntarily enter into this study, and those who would not.

## A Vision: Automatic, Privacy Protected Blind Utility Auditing

Ideally, Utility companies would participate in a third party system that would upload a different random sample of a region's housing (and/or commercial structures) from the county records each month. The system would have a link to utility company databases that would allow it to download utility use data from homes without disclosing any personal information about the occupants. This way, each month a new sample could be taken and automatically tabulated with the previous month's and over time, a very solid picture of energy and water use patterns would form. The information would not be linked to individuals, but to structures. This analytical tool could be replicated and used in any region to help define a baseline and performance goals while simultaneously tracking progress. Unfortunately, at this time the willingness and/or funding is not available to make such a system a reality.

## COST AND ENERGY SAVINGS FOR THE CITY OF TUCSON AND CIVANO

Cost for utilities per year and cost for heating and cooling per year were averaged (Figure 4.7). Cost savings are generally directly proportional to energy savings, but different utility rate structures can slightly alter this relationship. Additionally, behavior and even home design affect the relationship between savings and rate structure. For example, an off peak rate structure rewards energy consumers to use energy during times when there is lower demand on the grid, usually during the night time.

One technology that takes advantage of off peak energy is the ice storage air conditioning system. All night long, the consumer's air conditioning compressor uses less expensive, off peak energy, to make a large block of ice in an insulated container. The next day, the block of ice is used to transfer cool air into the home, without ever switching on the energy-consumptive air conditioner compressor. The overall process probably uses more energy than directly cooling the space would, but it uses cheaper energy.

Another great example of behavior meshing with home design is found with solar hot water units. If a person typically waits until midnight to wash clothes or shower, much of the heat gained from the sun that day could have dissipated. If the inhabitants choose to run their machines during the middle of the day, then they will be getting the full benefit of the solar water heaters. So certain behaviors do compound savings (or losses) of functional home designs.

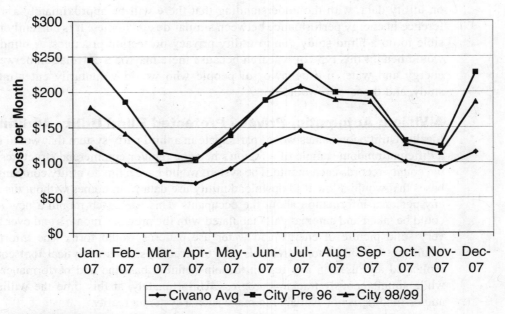

**Figure 4.7** Average cost for utilities and heating/cooling per year.

It is important to remember that these savings calculations are simple and do not take into consideration the compounding effects of savings. The coal-derived electricity that powers most of the state of Arizona comes at a heavy environmental cost. Every kilowatt hour that is saved reduces $CO_2$ emissions by 2.2 pounds and water consumption by 1/2 gallon at the power plant. Further savings could be inferred about the energy and water it took to mine and transport the coal. It also required electricity to pump water to the power generation facility and the mine in the first place, which again required some amount of coal to fuel that process as well. Following this line of thought leads one back to the complexities of understanding embodied energy, discussed later in this chapter.

# Water Use

Civano (per the MOU) adopted the 1998 Sustainable Energy Standard (SES) for water use as 28 gallons per day per capita exterior (landscaping) use and 53 gallons per day, per capita interior (domestic) use. Potable and reclaimed water are metered individually for Civano residences. Data from potable water use by 67 individual Civano residences and from reclaimed water use by 31 residences were supplied by Tucson Water Company. As part of the IMPACT process, it was determined that the 28 phase II homes in the study would not be required to deliver reclaimed water as an option on their homes (MOU, section III, A, 2). More will be explained about the IMPACT process throughout the book.

One year of water data was reported by Tucson Water Company for the city versus the Civano average. Samples returned indicate an overall average monthly potable water use as shown in Figure 4.8, the table below. These totals do not include the 15 Civano phase II homes that did not have a complete year of data. Tucson Water Company provided data from the total Tucson population of residential water users as 131 ccf per year compared to 92.16 per year (total water) for Civano I homes and 81.92 for Civano II homes. Civano I used an average of only 54 ccf/residence per year of potable water whereas Civano II's overall use is all potable at an average of 92 ccf/residence per year.

Thus, total Civano potable water use is significantly lower than Tucson homes—given the homes in the study. Overall water savings is likely a result of strict landscape standards, small lot sizes, use of rainwater collection cisterns, reclaimed water, and community awareness.

| Residential Water Use in Civano Compared to the City of Tucson at Large | | | |
|---|---|---|---|
| | Civano I | Civano II | City |
| 12 Month Potable Use (ccf) | 53.87 | 81.92 | 130.98 |
| 12 Month Reclaimed Use (ccf) | 38.29 | 0.00 | 0.00 |
| Total | 92.16 | 81.92 | 130.98 |
| Reduction Over City Average Use, Potable | 59% | 37% | Baseline |
| Reduction Over City Average Use, Total | 30% | 37% | |

**Figure 4.8** Overall average monthly potable water use.

This study year, site irrigation water was monitored for the Civano I neighborhood and yielded a result of 34.4 ccf/residence per year for site landscaping. The site is all irrigated with reclaimed water with the exception of the community pools which were not included in the calculation. The calculation does include the reclaimed water used to irrigate a large grass play field.

As seen in Figure 4.9, the city's average water use is higher overall than Civano, and the peak water use (June-July) is more pronounced in city homes as well.

As can be seen from the spread in home water usage, large variation characterizes homeowner behavior. Previous reports have indicated a similar range in use patterns. Figure 4.10 compares City of Tucson water use with Civano.

## CIVANO PHASE I WATER USE IN COMMON AREAS

Potable water is only used for the community pools. Elsewhere, reclaimed water is used for common area landscaping needs. In addition to the individual residential total and potable water savings shown here, the common area landscaping uses xeriscape (low water use landscaping) and reclaimed water, which further decreases usage of potable water while successfully providing shade and grass spaces in the community. See the ANE, Inc. 2001–2002 report for indications of the substantive contributions from use of reclaimed water for common areas (not computed this year as build out continues). A larger effort can be made to reduce water to native trees that are now established and stable by taking them off of any watering unless there are periods of drought during which supplemental water is needed for the plants' survival.

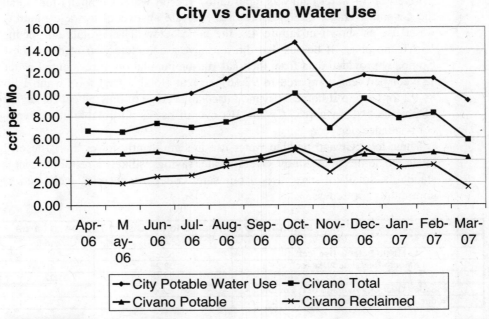

**Figure 4.9** Average water use.

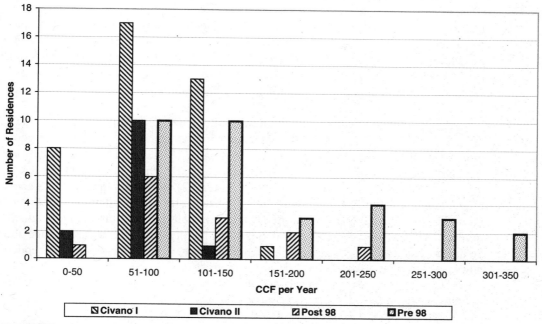

**Figure 4.10** Water use comparison.

## Civano Phase I vs. Phase II Comparison

Phase II of Civano is called Sierra Morado—homes designed and (being) built by Pulte Homes to meet the requirements of the SES and MOU. Pulte's vertically integrated company uses mostly in-house architects, planners, engineers, craftsman, mortgage brokers, and salespeople. Pulte is truly integrated to provide production homes. Phase I neighborhood design was master-planned by out of state and local architects and sold in small pieces to many different home builders. Thus, the way phase II is being developed differs enough from phase I to warrant direct comparisons of building pricing and performance. Although phase II lacks some of the character of phase I, it provides much more affordable housing that matches and exceeds the performance of phase I homes.

### MOU ADAPTATION

Phase I included connections at every home for city reclaimed water. This proved to be a large expense to the city and homeowners alike. In addition to an approximate

$3500 total cost to connect to the system (for a backflow preventer and a second water meter) the reclaimed water costs more than the lowest tier potable water cost that most Civano homeowners in our study pay. Reclaimed water is lower quality than potable and requires yearly maintenance of backflow protectors. This added cost and maintenance has created an incentive for homeowners to abandon or not use the city reclaimed system in favor of other methods such as rain water collection. Considering the huge cost of the added infrastructure and low incentives to use the system, city reclaimed water is not appropriate for use on the residential scale unless it is made more financially attractive.

With these new lessons learned, the neighborhood decided to drop the requirement for city reclaimed water in phase II and III at the individual home level; however, common areas must use reclaimed water for landscape. Homes will use low water use landscaping with berms and swales to enhance growth. Phase II (unit 1) has 12.45 acres of landscaped common area and phase III (unit 2) will build out with 17.23 acres of landscaped common area, all of which is irrigated with reclaimed water. The homes on traditional lots don't have reclaimed stubbed to their property, but the high density product, the landscaped right of ways, the community center landscaping, and the common areas are all served by reclaimed water. When one considers the embedded power in water, the costs associated with the infrastructure, transport, and waste treatment, delivering reclaimed water is not always as practical a solution as it may seem at first consideration.

# Home Resale Values: Phase I, Phase II, and a Neighboring Development

*The term 'sustainable development' has been defined as 'a process of change in which the use of resources, the direction of investment, the orientation of technological development, and institutional change all enhance potential to meet human needs both today and tomorrow.'*

—Civano Impact System Revised Memorandum of Understanding on Implementation and Monitoring Process, December 8, 2003.

As development has progressed in Civano there has always been the question of how much it costs to adhere to the advanced requirements of the SES and MOU. This has also been a major concern for developers. Now that phase I has sold and homes in phase II are being priced, one can begin to make the comparison. In a generalized analysis of MLS information from 18 Civano phase I homes, eight phase II homes, and 17 traditional development homes (non SES/MOU) just to the south of the Civano development, the following trends are beginning to take shape:

1. Given the MLS sample of homes sold between 7/26/07 and 1/22/08, Civano phase II generally sold for 13 percent less per square foot than homes in a "traditional" (non SES/MOU) production housing development.
2. Civano phase I generally sold for 33 percent more per square foot than homes in phase II. In comparison to a nearby traditional production development, Civano's first neighborhood was approximately 20 percent more expensive. Cost per square foot data is not consistent. Access to information that would make the results of this comparison clearer, such as builders' profit margins, lot sizes of homes sold, etc., are not available. Some homes have more or less attractive locations within the developments, larger houses tend to cost less per square foot than smaller ones and different homes have different features and more data will be available as Pulte sells more phase II homes. Be that as it may, the trend is clear—meeting higher performance building standards does not demand high cost housing. Civano phase II is subject to all of the same energy standards as phase I but is currently very competitive in terms of costs—approximately 13 percent less than even normal homes in the same region given this sample. The Urban Design of phase II is simpler than phase I, yet still incorporates many of the features that make it a walkable community with shared open spaces and recreation spots.

What accounts for the extra 20 percent premium Civano phase I costs? The neighborhood center? The extra effort put into neighborhood design and street layout? Time and study may answer these questions. However, even after the numbers are in it may be a mute point. Fannie Mae subsidized millions of dollars worth of development in phase I that were never fully passed onto home buyers. Remember:

"The goal of the Civano project is to create a mixed-use community that attains the highest feasible standards of sustainability, resource conservation, and development of Arizona's most abundant energy resource—solar—so that it becomes an international model for sustainable growth." —Civano Impact System Revised Memorandum of Understanding on Implementation and Monitoring Process (Appendix A).

Phase II prices include actual costs for a profitable project that may be repeated in the future—something that cannot be said for phase I. (See appendix for chart)

## Inter-Civano Energy Comparison

Overall energy consumption in phase II housing (Pulte Homes) has increased over phase I. This may be due to the small size of the phase II sample, behavior, building quality, or sampling methods—but for now this difference is non-conclusive. Pulte includes the release form for this utility study in its signing documents that presents a difference in who signs up for this study. All other releases are solicited in a 'cold sale' fashion and it may be that the individuals responding are more apt to be aware of the effects of their behavior on energy and water consumption than those who are agreeing to the study as a matter of course to completing their home purchase.

In the end, human behavior is the missing link to this study and perhaps the key to becoming a more sustainable society. It would be extremely advantageous to conduct a social-science study to bolster the findings of this (and any other) residential energy/water audit.

# Inter-Civano Water Comparison

There are some very apparent differences between the main phases of Civano's residential development. Phase I was built by many different builders and was under the management of different master developers. Phase II has been entirely built under one company, Pulte Homes, with Pulte as master developer. Thus, the business model under which these later homes were built was different. As for the adaptations to the MOU and SES via the IMPACT process, phase II does not gain potable water savings from the use of city reclaimed water because those houses are not connected to the city-reclaimed system. Therefore, phase II uses more potable water than phase I. However, the *overall* water consumption in phase II was lower than phase I by nearly 11 percent during the timeframe of this year's audit (see Figure 4.11).

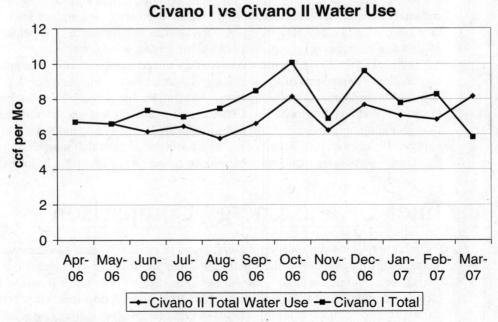

**Figure 4.11** Water use comparison for Civano phase I and phase II.

The retrofit of inexpensive and uncomplicated rainwater harvesting systems may be an excellent way to reduce potable water usage in further Civano homes, as many homeowners have already done. These systems only cost approximately $1000 less than the connection fee for city reclaimed water, are nearly free to operate, and do not require the extensive infrastructure that is needed to redistribute reclaimed water. Using grey water is possible, but more problematic, requiring excellent homeowner education and willingness to maintain the system. It is very inexpensive to incorporate grey water stub-outs into each home during design so that grey water may be utilized by interested homeowners. (See Chapter 7 for more about building and water technologies)

The comparisons between the performance of Civano phase I homes and Pulte's phase II homes show slight lagging in energy efficiency of the phase II homes over phase I. Phase II showed an overall decrease in water usage of the phase II homes. Phase II homes use more potable water than phase I homes because phase II is not connected to city reclaimed water at the individual home level. The phase I landscape also has much more foliage and a larger overall landscaping water budget. However, phase II homes are meeting a project goal that phase I has largely not met: increased affordability.

The process through which the goals for Civano's energy and water reduction targets were identified and are measured did come with a price tag. Oil over charge restitution funds, grants, countless volunteer hours, and grant money helped with defining strategies and goals. Now, yearly costs for the audits are modest, and are shared between the master developer (now Pulte Homes) and the City of Tucson; further linking the private and public effort that typifies Civano. But what are the costs of not measuring? It would be the equivalent of making a goal to drive somewhere without a clear idea of either the starting point or destination—perhaps a nice way to spend a vacation, but a lousy way to get somewhere specific.

# LEED—Neighborhood Development "Certification of a Completed Neighborhood Development"

The public and private partnership that guided the design development and monitoring for Civano has proven a thorough, yet cumbersome process. It came at a time when there was less data about the resource savings of specific high performance building methods and before mainstream high performance certification programs. Now, there is a quickly growing body of knowledge about the synergies of high performance design and a quickening transformation of the industry unlike ever before. Therefore, programs tend to be front-loaded, focusing on the integration of design elements known to increase performance and measured by how many such elements are incorporated into the final product.

Certification for a LEED project, for example, requires documentation of every design strategy that the U.S. Green Building Counsel has been able to determine will improve the environmental performance of a structure or neighborhood. Performance strategies each have a potential point value and a certain number of points are required to reach a given level of LEED certification. What makes this process stand up is that performance strategies are scrutinized by a third party USGBC committee unconnected to the subject project. Strategies that are deemed appropriate to the development scenario are assigned points that can count toward final certification. Not until the entire project is built and every point is verified does a project become LEED certified.

Unlike Civano, LEED projects are then assumed to be operating at a higher performance than status quo projects. Verification of performance via an audit study is not required, only the verification that original design and methodologies for attaining LEED points were completed must be documented. The LEED method is intrinsically different from the process at Civano. LEED certification combines as many synergistic high-performance design elements into a project as possible in order to attain the highest rating possible, whereas with Civano the goal was to achieve a specific increase in performance above and beyond something already known whether it takes one performance design element or one thousand. Therein is the fundamental difference between prescriptive approaches to sustainability and performance-based approaches. Civano's process mandates that the performance be checked, which makes sense given the fact it is a performance-based system.

This is not to say that there are not audits performed on LEED certified buildings. Quite the opposite is true. It is to the benefit of those professionals who use the LEED process to know how their prescriptive approaches have been working and to be able to describe the successes in real, quantifiable terms. Often, those investing in high-performance design do so as a conscientious business decision and, as such, they monitor their return on investment by closely watching their utility bills—a business choice they made, not any kind of rule they followed. The USGBC's great strength is its membership base. They bring their own experiences to the table and help ensure that the foundation of knowledge behind the LEED system prescribes methods that will yield results.

Thus, the key difference between what is experienced in Civano and what is offered by the Leadership in Energy and Environmental Design certification program is choice. Civano is an example about the stick and LEED is about the carrot. Civano changed the zoning and development law in a small, specific region of Tucson as a test bed for high-performance codes and came with all the complications of enactment, enforcement, and measurement. If somebody wants to build there, they *must* follow the development guidelines, even though they are above the minimum code requirements for the rest of the city. LEED offers the promise of designers knowledgeable of a robust methodology for providing enthusiastic clients with high-performance projects and the reassurance of the U.S. Green Building Council's seal of approval—but participation is optional. So, which is better—the carrot or the stick?

Well, that is increasingly complicated. Although this book speaks more about LEED processes, it is because the LEED—Neighborhood Development pilot program is the first of its kind, a neighborhood development program that could be used to develop a project on the scale of Civano. But there are countless other high-performance building programs across the country and several in the state of Arizona alone. The National Association of Home Builders (NAHB) published green building guidelines many years ago—guidelines that helped shape Pima County's green building program. Scottsdale brought forth Arizona's first official 'green' building program and currently the City of Tucson, located inside Pima County, is formulating its own green building program. For everybody involved, keeping afloat with the standard evolution of baseline building codes can be dizzying.

Civano shows that a system can be devised that can raise the bar on *all* builders, designers, and jurisdictions of authority across the board. The public and private partnership yielded a performance-based efficiency standard that has proven attainable including a method to monitor and adjust the standard to change with the times. Green building programs give consumers choices and professionals the tools to deliver. However, the ideal situation would be that a time comes when green building programs become mute because green building is the standard and expected practice. What we see today is essentially a race to see which green building program, or programs, will be positioned to become the standard building codes of the future.

From the viewpoint of an engineer, the degree of success (greeness) comes down to measurement. As supporters of the U.S. Green Building Council and the LEED program, the authors of this book recognize the benefits of LEED certified structures. Furthermore, having been chairperson and served with the sub-committee that helped amend the NAHB's green building guidelines to become Pima County's regionalized green building guidelines, the authors also support municipalities that internalize the green building doctrine and offer benefits to those using these high-performance techniques within the jurisdiction's boundaries—however confusing it may become if they all do so. However, what none of these programs provide as an integral component of their program or process is a means to determine the level of performance being attained for a given amount of effort—independent studies aside.

## EMBODIED ENERGY

Embodied energy is fundamental to the understanding of performance and so exploring embodied energy sheds light on the complexity of the challenges to measuring sustainability. Embodied (or embedded) energy is the sum total of all the energy required to provide a good or service, from the raw material extraction, to transport, manufacture, and delivery to the point of use. Calculations for determining embodied energy of a specific material or product as a singular value is not possible. For example, if every single piece of PVC piping came from the same raw material resource and was manufactured in the same plant, the embodied energy of each unit of PVC pipe would be fairly consistent. However, the piping must be shipped around the world by different means, in different quantities, in vehicles of varying efficiencies—which means that individual loads of PVC piping will have different embodied energies.

Furthermore, raw materials are extracted and manufactured into products in places around the world. Different mines and factories use different vintage technology and methodologies. Therefore, a PVC pipe manufactured in one location may have a completely different embodied energy value than one manufactured in another, even before shipping. Each component of the factories building PVC pipe has an embodied energy as well. As each component serves out its life, some number of feet of pipe will be created before the component wears out and must be replaced. Thus, part of the embodied energy of the pipe includes the energy required to harvest the raw materials, build, deliver, and install the components of the pipe factory. Accurate embodied energy calculations become very complicated, well beyond the parameters of this simplified example.

The story of energy in a product, such as PVC, is a chain of events that goes beyond any practical application in terms of high performance building design, but does serve to tie concerns of embodied energy directly to the environment. If one backtracks down the chain of events resulting with the final, delivered PVC product to the earliest events and energy inputs that made that product possible, they find themselves earlier in time than the first life on earth. This may be an extreme statement, but the energy embodied in oil, coal, and other long-cycle hydrocarbons took millennia of energy input from our sun and the gravity of the earth to become the refined resources we are using now.

Embodied energy in our mechanized world is the combination of human activity and energetic reactions of fuels that have stored energy from nature. It took people to utilize the fuel—which means that the source energy for our industrial world, every bit of the energy that created machines to harness fuels, was started in motion at the direction of our ancestors digging in the dirt. The web of connections between the energy flows required to create something in the industrialized world is dizzying. It is not something that is easily calculable, but could be more so with much higher levels of business transparency.

So, although it is not yet a mainstream measurement, embodied energy of products and processes is a measurement we should research and develop to better understand terms of efficiency, just as we do our building envelopes. If one day businesses are able to operate with much greater transparency and attention to energy details, it would be possible to devise a consumer rating index that would clearly convey an idea about the energy embodied in a product delivered by a given process, sometimes referred to as a sustainability index.

This way, companies that improved upon their own processes could be recognized as such and incentives to do better would encourage companies to seek down-line companies that are more efficient than the down line of their competitors. One must also consider externalities and whether a product is recyclable or not. In a perfectly performance-based future world, companies would operate with full transparency and provide a consumer label on all products, including homes, which would give consumers a clearer understanding of the embodied energy implications of their purchase compared to other choices.

# A Closer Vision

The measurement of green as a performance target is coming about. Increasingly, high-performance projects are monitored post-completion and compared to similar status-quo projects. The resulting data is helping designers hone their art and maximize the synergies of their performance strategies. Furthermore, regionalization is becoming more and more prevalent—a very good reason why different jurisdictions may need to adopt different variations of a green building program, if not create entirely unique programs. The ultimate goal is that during this adolescent time in the world of high-performance development we determine what works and what does not via the diversity of methods popping up across the country today; and then we turn the very best ones into our zoning and building code laws of tomorrow.

Lastly, we would measure the successes and failures of our efforts and continually push the bar ahead of the baseline measurement using a system akin to Civano's IMPACT process. Funding that currently goes toward government green building programs could be reallocated in the future to pay for the type of study that is carried out yearly for Civano—the results of which could inform the zoning and code development for the region in question. Professionals certified in the LEED process, familiar with NAHB, or other green building systems alike, would all have the basic training and knowledge required to continue providing the most environmentally conscientious development planning that their customers can afford.

# 5

# GERMINATION

*Let us develop the resources of our land, call forth its powers, build up its institutions, promote all its great interests, and see whether we also, in our day and generation, may not perform something worthy to be remembered.*

—Daniel Webster

The overwhelming reasoning in support of land development is to provide for the needs of a growing population. The modern and healthy lifestyle to which we are accustomed requires housing, farms, transport, schools, hospitals, churches, grocers, and all means of basic services. However, the requirements for developing an untouched patch of land into a village equipped, in part, with all aspects of human necessity is neither simple nor basic—and ever less so as our environmental perspective widens. As with any complex task or project, there are often as many methodologies to achieve completion as there are people working on the job. Consequently, multiple ideologies follow suit and questions abound; though two primary questions consistently rise to the top:

**1** What are our goals?
**2** What is our budget?

Many people reading this book have been to design charettes or have participated in LEED projects. All have been in an organized work group. Less potentially impactful projects still evoke friction, even when nobody on the team feels like their choices will affect global warming, resource depletion or long term quality of life for their children. More often than not, land development issues at large are becoming the crux of such passionate concerns. There are those who find an intrinsic motivation to do better (build green) and decrease environmental impact. And there are those that realize there is money to be made in a changing market. Most in the green building industry are hybrids—people who know there is a living to be made in transitioning to doing the right thing.

This chapter is about decisions. Civano had moved passed being a mere seed and achieved germination. As it blossomed, infusions of new ideas created a depth of scope that was very complex, controversial, and challenging. As a result, development companies bought and sold the project, effectively transplanting the seedling of

Civano (the land use agreement, zoning, etc.) between several different hosts. Each host had a different point of view, a different set of constraints, and a different answer for the two questions above. Thus, each left their mark on Civano in different ways. Interviews with project participants will illuminate different points of view across the gambit and help tell the story about why and how different choices were made for Civano—and expose lessons learned. Refer to Figure 5.1 for an overview of Civano's final layout design.

Al Nichols' interview with Lee Rayburn:

**Nichols:** When did you first hear about Civano?

**Rayburn:** I first heard about something called the Tucson Solar Community, probably in 1985 ... and I thought, "This is pretty cool stuff, but it makes no economic sense to me at all." So I just sort of tracked it as something way off in the distance. The next time I really had anything to do with Civano was when the original planning charette happened in September 1996. A friend who lived downtown offered to let me house sit and I figured, why not? And a friend working for the city, Doug Crockett, said, "Oh, by the way, they are doing the design charette for Civano, you should go." I didn't even know what a design charette was, and there were these guys, Stefanos Polyzoides? Andres Duany? And so I came out! And for a day I hung on the sidelines of the charette and for the first time met Stephanos, Andres, John Laswick, and I sort of got intrigued by it. So I came out the second day and was just sort of observing it all.

But on the whole plane ride back to Baltimore, where I was living and working at the time, I kept thinking about the charrette and the master plan that was emerging from it. I couldn't let go of thinking to myself, "They are making a terrible mistake where they are putting the neighborhood center. It really should be down by the main road ... Houghton Road." But I told myself, "Don't worry about it Lee, because you will never have anything to do with Civano again." Funny story about how sometimes life goes ... never assume anything.

Then, in 1996, I had made a decision to let my little architectural business in Baltimore just sort of wind down—I needed to move and I was looking into Houston, primarily. But no matter what I looked at, what jobs I considered, I always kept getting guided back to Civano—to Tucson. And in the summer of '97 I came out and interviewed for a job with David Case and Kevin Kelly and we hit it off. And the rest is history. So I came out here in August of '97. My job description at that time was basically, "Please, make it work!"

**Nichols:** Was this before or after the auction?

**Rayburn :** This was after. This was after David Butterfield had gotten out.

**Nichols:** And why did he get out? Did he get common sense and decide it couldn't work either, or?

**Rayburn:** No, I think it was more to a necessary business decision on Butterfield's part. He had stepped up, wanted the land and the project; and really

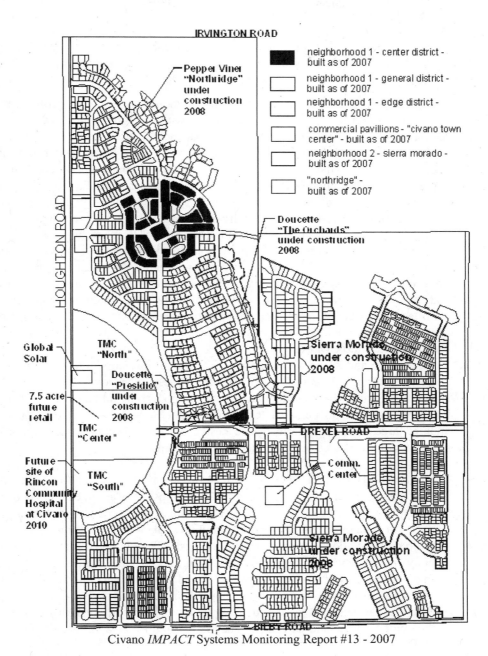

**Figure 5.1** Civano site map, all phases.

got it started in the right direction. I think at the time of the auction he approached David Case and Kevin Kelly and they formed a partnership. And within a year, or less than that, it became clear that more capital would be needed than Butterfield could come up with. If memory serves, David Butterfield then decided to hand the entire project over to Case Enterprises, which is to say David Case and Kevin Kelly.

Just like any investment, land agreements are bought and sold for profit and sometimes, at a loss. The land happens to go along with the agreement. Civano was more than an agreement, it was a push forward that incited shock and awe with investors beginning with the first major (post master plan) investor, David Butterfield. After it became clear that the Civano agreement would be potentially very risky, Mr. Butterfield made a sound business decision—to turn a potential liability into profit by selling it. Civano had overreaching intrinsic values associated with it, and the sale seemed a callous decision to some individuals invested in other ways than only monetarily. However, Civano was a frightening investment to make. It came with all the risks that early innovators take and there was already a profitable business model for the residential construction industry. Civano was being built on ideals; and because both authors of this book have individually failed at least once in an idealistic business endeavor, we empathize with how difficult it can be to sell ideology. But then, there are often those who will buy-in for various other reasons. David Butterfield's partners, David Case and Kevin Kelly, bought Butterfield out in early 1996 and in doing so, bought further into Civano.

David Case and Kevin Kelly had a basic group of investors they had built partnerships with during operation of their earlier business endeavors. Their story, and the story of all business entities associated with Civano, is only told here to the best of our knowledge and in no personal terms. Their story can help other developers who attempt large scale high performance design be more successful, save money, and make high performance building increasingly affordable.

**Rayburn:** David and Kevin had great success prior to Civano and they had long-standing and very good investor relationships. They assumed that once they really figured out the Civano project, they could go back to their old partners and say, "We have this great new idea, we think it's fantastic, let's just continue with the type of relationship we've had in the past." They bumped into what everybody would have bumped into back then, which was having their biggest investor going, "We love you guys, we've had great partnerships together, when you're doing Civano number three, come back and talk to us." Civano at that time was basically just too cutting edge for them. As institutional investors, I bet they didn't understand half the language Case and Kelly were using to describe all the goals of Civano.

In almost every situation the lessons learned in Civano come back to a question of education. Education of the investors, the developers, potential buyers, the authority having jurisdiction, and the builders was a large portion of the Civano task. But starting with something new often means that those initiating the process are being educated as they move along as well. This is another reason why early innovators often meet with struggles and why investors are often fickle in subsidizing unproven innovative efforts.

# Three Major Lessons Learned

## #1 LESSON LEARNED: CAPITALIZATION

Bob Cook talking to Brian McDonald about Civano business:

**Cook:** Well so you know that there have been many businesses that sort of came and went. They just haven't been able to be sustained.

**McDonald:** Do you know anything about why that is?

**Cook:** Well, I think there just wasn't enough traffic. There was a coffeehouse for awhile and now it's moved away and into Safeway a few miles down the road. When we heard that Fannie Mae was coming in, of course, we were really happy because we knew that Case was having problems and—in fact, when Case came in, we were really happy because finally we found a private developer that was really interested in the vision of the project and wanted to be the developer of the project.

**McDonald:** Do you feel that sentiment would hold true, later on down the road?

**Cook:** Well, I don't think David Case and Kevin Kelly had any particular experience in this type of development. They'd done a lot of multi-family housing and low income, affordable houses, entry-level housing developments. And so they had experience in developing housing. But they had no experience in the breadth of a project like this that had energy and water use goals and employment and environmental standards that no other development has ever been forced to adhere to in this area.

That's why this was a pioneering project, because Tucson had never had this type of development occur before. It was public-private partnership, which meant that the original vision of it was developed in a public process with public funds, if you call those Exxon overcharge funds public funds. They were private in a sense that even though it was our money, it wasn't tax money. So, there should have definitely been a benefit to the community. And that's why we thought that this thing should be designed with the highest standards possible, pushing the possibilities of what could be contained within what's called a sustainable model community.

In fact, when this advertisement poster was done, I think was one of the few first times the word sustainable was ever connected with community or residential development. And so it looked like Case was having some challenges—there were costs incurred that were probably unnecessary had it been better thought out. So I mean, these kinds of things added up. There were issues with the grading expense to achieve drainage and the street layout. There was the added cost of reclaimed water. There was misunderstanding about high-tension lines. Some people thought the lines would be buried ... a perceptual misunderstanding because TEP knew which lines they'd bury and which ones were ridiculous to bury.

But in any case, after a couple of years, I think they wanted out, you know, and they found someone to take it over. And I don't know the story of who found Fannie Mae or how Fannie Mae got involved. But there was a lot of hoopla when groundbreaking occurred when we had our first public celebration of this development.

Councilmember Shirley Scott spoke at it, as did Jane Hull, the governor at that time. Officials of Fannie Mae were also part of it.

It was the Fannie Mae money that was able to fund the full infrastructure and initial community buildings. They had the deep pockets to do that in a timely way. And so it seemed that, you know, things were maybe moving forward. But little did we know that Fannie Mae was, from a management perspective, just the wrong entity for this sort of thing.

One lesson learned during the course of developing Civano is that early financial preparation is increasingly necessary the more integrated and/or complicated the project. Top priority in developing a new design idea is pre-developing the notion with likely investors. Investors are generally more likely to put in for what has been successful than something they know nothing about. There are naturally occurring difficulties normally associated with transforming an organization. Most of these difficulties are due to the fact that people generally fear unknown quantities and so by nature we usually tend to resist change.

"Resistant to change" has been an understatement in the real-life industry setting because the organization in question is a megalopolis builder. This organization is a machine with incredible inertia and a tendency toward sameness, codes, inspections, regulations, jurisdictions, plan reviews, return on investment, and a love/hate relationship with accuracy of performance. This organization needs a whole infrastructure of its own to operate. It neglects to change without strict enforcement or a simple need for business survival. The conflicting interactions between codes enforcement and business survival are the new fuel toward advancing building science and development innovations.

New building codes are predominately predicated by accidents. When a building burns, officials figure out why to the best of their knowledge. Then they decide how to improve the building codes in ways that will limit the changes of a similar accident from happening again and ways that will diminish the impacts of the accident if it does happen. For example, some time ago balloon framing was banned from the codes because the process, although quick, left chimney-like voids between levels of a building which were perfect channels for fire and smoke to travel through at an alarming rate. Also, there are mandatory egress rules that must be met in all permitted buildings to best ensure occupant escape in case of a fire.

Thus, over the years, code after code, timber framing developed into a routine process that, for much production housing, has become the predictable standard. The code inspectors and most builders alike know the rules and they check for double plates and properly framed stairs. They look for correct nailing schedules, material usages, stud spacing, and blocking requirements. They follow an established formula for success. Framed buildings built to code standards are considered safe dwellings. They do not pose an immediate threat of any kind to their occupants under normal operating conditions. There is little need to address the code for stick framing then, right?

Enter the new age of green and sustainable concerns. Consider this: a traditionally framed home will consume approximately 25 percent more lumber than a home framed using what has been called advanced framing techniques. So, should advanced framing techniques be required by code? Why would a retroactive code system care? Although advanced framing techniques are not required, codes are now changing rapidly and in a direction that begins to view the whole life cycle and environmental cost of a building as part and parcel to the health and safety of people, and thus a valid code concern (although advanced framing techniques are not mandatory). So codes are increasingly becoming proactive and in the 2006 cycle of the International Energy Conservation Code (IECC) the definition of a compliant building increased approximately 10–15 percent in efficiency.

At the time of this writing, the 2009 cycle of the International Energy Conservation Code (IECC) has been completed and is sufficient to increase building efficiency across the board by approximately an additional 15 percent when compared to the 2006 IECC. The code is expected to jump another 15 percent or more in 2012, which will bring buildings built under the 2012 code to a higher performance level than Civano buildings in terms of envelope thermal properties. The code will also require solar stub-outs for simplifying potential future connection of photovoltaic and water heating apparatus.

There is usually a several year lag after codes are created before actual practice follows, assuming enforcement is in place. If the Tucson City Council adopts the 2012 international codes, there will be a vetting process whereby code committees make adaptations to the code language for adoption. This process may take up to a year. To add to the slow uptake of new codes, model home plans are generally exempt for some time after codes are adopted. In Tucson, permitted model plans can wait up to three years to adapt. So, as a general rule of thumb, one can assume that even after codes are developed and announced, it is at least two full years before true change can start to be seen in the way homes are built.

For perspective, during the first half of Civano's lifespan from conception as the Solar Village in 1982 through 1995, no major energy code changes occurred in the region. During the next 17 years, from 1995 until 2012 code energy code requirements will incrementally increase to accrue approximately 40 percent in envelope efficiency gains. Even though it may be 2014 or 2015 before the effect of the 2012 increase takes hold, this is quick movement from the behemoth land development industry and when coupled with the green revolution, there is a new mentality of embracing change coming from the building industry.

Other considerations about code adaptation have to do with regional climate concerns. Until recently, energy calculations predominately used heating degree days which reflect the demand for energy needed to heat a home or business. This makes sense in much of the United States where heating loads often outweigh cooling energy requirements. In Tucson, most of the land is below 4000 feet altitude where there are only about 1400 heating degree days and much closer to 7000 cooling degree days. As mentioned in Chapter 4, the original model energy code was modified to become the SES—requiring that buildings in Civano meet a 7000 heating-degree standard. Now, energy compliance modeling programs and energy code language is beginning to utilize cooling and heating degree days for accuracy given the geographic zone and climate a project is located.

When so many entities from the grand industry of land development are affected by code rulings, they generally are eager to influence how those codes are decided. They have volunteer groups, lobbying groups, and special interest groups all seeking to protect their interests—whether their interests are their neighbors' well-being, their bottom-line earnings, or a little of each. In this context, it becomes easier to see why proactive change can be met with such resistance. Also, in this context it becomes clearer that the role of everyday citizens in guiding land development practices is realized through participation—showing up, offering perspective, and attempting to balance the often purely business-driven interests of some large developers.

## #2 LESSON LEARNED: SIZE MATTERS (CAPITALIZATION CONTINUED)

**Rayburn:** And this was a big, big project. You have to be someone like a Pulte Homes to take on 800 or 1200 acres. To do a project of Civano's size, you really have to be a pretty big company because it requires more capital. The situation that Case Enterprises got into was that of a small, lightly capitalized company with a gigantic interest clock ticking just on the land alone. When I came on board in August 1997 you could almost hear that clock ticking every day. We constantly had to think about raising enough capital for immediate future because of the interest costs and the overhead and everything else. So, size matters, especially when you are trying to do something innovative and you really just have to be really smart about choosing the scale of the project because the scale can overwhelm you and you can end up spending far too much time managing your debt.

In my gut, I just knew that if you were trying to do new urbanist or neo-traditional planning, that there was a significant component of additional management cost in it.

To this day, there are signs that advertising project as green is dubious for some developers. High performance development not only seems more expensive intuitively, but it generally is more expensive in terms of up-front planning costs as well as in construction—but this doesn't speak to value. Civano, with the help of Pulte Homes, has shown that environmental considerations like home energy and water efficiency can be built into production housing at a very low additional cost. Many developers offer a great number of energy and comfort upgrades standard when compared to code. One may wonder why more do not take part in the green marketing feeding-frenzy. Some say that a marketing scheme for green building is like a marketing scheme for organic food. Everybody knows organic is better in many ways, sure, but it is usually darn expensive. The conception that good equals expensive is a tough marketing roadblock to overcome.

**Rayburn:** The other thing that was really apparent was that this [Civano] was a program that had its genesis in, and was being really advocated for, by a group of activists. If you have a development program being promoted by activists, you have to look very deeply because, typically, it's not going to be about the money, it's about the ideas.

**Nichols:** And money be damned.

**Rayburn:** Yes, and money be damned. And in fact, after I came out here and started to stumble on the old documents, there was an article in the Tucson weekly during the mid- 80s when the idea of a Solar Village first came up. The lot values were going to be like $200,000 per lot. Again, nobody thought about what that meant because they were so entranced with the idea. Who was going to buy those lots?

The message about what Civano was became mixed and, consequentially, there were disagreements. Was it a niche development aimed at a small group of prospective homebuyers, like Wayne Moody's Milagro Cohousing project, or a new mainstream product that could effectively be marketed to a wider cohort of homebuyers? These questions were being asked mostly well after the scale and master plan of the development had already been decided and the land auctioned to the first developers. The question of scale is an integral component to the question of marketing. Simply, how many people are there who will buy into a given idea or product? Market demand then will inform the quantity of units that should be produced.

Generally, a much larger scale project needs to be closer to mainstream to find a large enough market segment. A smaller scale project can afford to specialize more, at the cost of reducing the pool of prospective consumers. The public and private partnership process that led to Civano started with the land selection and a concept that the place built would have some on-site jobs and be powered by solar energy, but no attempts were made to identify how many people in the area would actually purchase such a product before the site was chosen. From the onset, a large parcel was sought and the final choice encompassed 820 acres. Thus, Civano's scale was largely decided but was not fully considered as an integrated component of project specialization.

Bob Cook on the Village Homes project in Davis, California as a model for Civano:

**Cook:** I had a friend that moved there, and I visited that project. To me, that is one of the great models because it was conceived entirely by a private developer who had to go through a lot of hoops and waivers and exceptions. And he had to convince the finance community as well as the bureaucrats and the city that this was a good thing. And he persisted and was able to build this community. And it was—it was so successful that there have been waiting lists for decades to either rent or to buy there—so it's been enormously successful.

That model, which was a much smaller scale, it was only 75 acres, included a lot of things which have become commonplace in talking about sustainable communities. It was a walkable community, the streets were narrow and they were shaded. The idea was to have the community interaction happen along the common pathways and the garden centers between the houses. The community gardens and solar houses and common areas enhance the interaction of the community and minimize the role of cars—making it a very people-friendly development.

And they even dealt with affordability issues. There were some homes that were built but not completed, and were allowed to be completed by the sweat equity of the buyers. So that allowed people to move in and do their own plastering and painting and some of the other finishes.

One of the earliest marketing phrases for Civano stated it was a place for "Poets, Thinkers, and Dreamers." This is a quintessential example of idea-driven marketing wherein a new, untested idea is presented to an established market with the hopes that the idea itself will be strong enough to attract buyers. In other words, idea-driven marketing tries to adapt consumers to the product. Market driven marketing takes a look at what people already like to buy and adapts the product and marketing message to fit the market.

## #3 LESSON LEARNED: IDENTIFY GOALS AND SEND A CLEAR MESSAGE

**Rayburn:** So, people had very mixed up ideas about what we were trying to do. Be careful about how complex your message is and then be very disciplined about a hierarchy within that message. It takes discipline to remember that "these items are the most important, but items three, four, and five are negotiable."

I think another big story of Civano was that even when all this started to happen, the concept of sustainability was one that was just starting to become much more public and be discussed. We were all beginning to think through ideas more wholistically. One example: We were looking at this cool composting toilet from Australia that does all these neat things. But then I was at a conference where someone started talking about the costs and impacts of transportation being part of the emerging discussion about sustainability. Oh shoot! Now, you know, you had to go back and rethink the toilets. We were so out there on the leading edge that we had to learn our own language internally, and it would change a lot, and then we had to go to the public and speak the consistent language to them.

Today language is still a large part of the challenge. This very book that elaborates about the confusion buzzwords like green and sustainable create still uses these terms often. Organizations like the USGBC and the NAHB have spent countless volunteer hours and funds to develop a language about environmentally friendly building that can be as concise and user-friendly as possible. The concepts in building science that must be explained are not all intuitive. For example, some companies market their product's insulation properties in terms of a steady state R-value while others market a dynamic R-value. How will the consumer know what the difference is? The term "low-e" has been attributed to windows in a generic way. Consumers are beginning to use the language "low-e" but they sometimes think this means insulation when in fact it is a glazing property that refers to infrared emissivity (or lack thereof). Ideally, glazing would be applied differently for areas where solar gain would be used for heating than areas where solar rejection is desired for solar rejection (for more building science information see Chapter 7).

Taken apart from building science, language for new urbanism can seem even more befuddling, not yet as grounded in math and formulae as efficiency ratings on HVAC machinery or window R-values. This confusion often extends to the same design professionals and builders who have a very strong handle on code and building science languages. Comparatively, the new urbanism movement is young and although it does have an aspect of intuitive meaning about how we relate to the places where we live and how we perceive our built environment, the meaning is neither the same for everyone nor can all the same concepts be applied in every situation. Therefore, it becomes increasingly important, and it is increasingly common, for community-scale projects to be guided by the language of a plan book or pattern book.

**Rayburn:** Let's back up to October '97. We had to get the tentative plat passed by city and council or the project would have died. Getting that done was my job. Also at that time David Case was saying, "We need to get an equity partner. We're not that far away from bulldozers moving dirt on the site. We have to find a substantial equity partner." Fannie Mae is a peculiar institution: chartered by the federal government, but acting as a private company. Its charter required it to do certain things, one of which was to encourage innovation in homebuilding. They had a special fund focused on innovative housing. They and we sort of bumped into one another. We both had an epiphany: "Wait a minute, we're [the Civano development team] innovation-oriented housing developers and you're Fannie Mae looking for innovative housing. This looks like a marriage made in heaven, let's make this happen."

So that fall was consumed by our negotiations with Fannie Mae. At the time it seemed like a gift directly from heaven. They certainly had the funding required for the project, and they said they'd support us. As we got down to the very end and time to sign on the dotted line, we got this last minute direction—and this is sort of simplified but it gets to the point. They said, "We love this project; we love you; but here are the things you can't do:

**1** You can never use the word "sustainability"

**2** You can never use the words "new urbanism"

**3** Other than that, go out and "sell, sell, sell!"

We looked at that and went, "Have you not been here?" I tell you this story because it is probably the clearest example of good news about finding a financial partner balanced by the realization that we were worlds apart in understanding the core concepts of Civano.

And that was sort of the end result of this intense search for money. It led to a financial marriage that had at its core deeply, deeply contradictory issues going on.

The conflict in vision between Fannie Mae who was invested in the older way of conducting business and the innovative approach of Case Enterprises eventually led to a powerful, larger company taking over the project as master developer and edging out David Case, Kevin Kelly, and eventually Lee Rayburn. This kind of thing can happen all too easily, but a very clear, concise, and consistent message and use of language can go a long way toward avoiding unknown ideological gaps between project partners. If the developers do not have a clear message, how will they communicate with their financial partners or potential customers?

**Nichols:** So did Fannie Mae decide the vision just wasn't working and go into recovery mode? It seemed like they changed from seeking the vision of their mandate, to banker—like they had a mission, and they just gave up on it.

**Rayburn:** Well, it is more complex than that. I used to go to a lot of conferences for my job and I met a lot of other people who worked with Fannie Mae. And I don't want to knock Fannie Mae because we would not be sitting in this mixed-use home and office, on this land, if it were not for the money they brought in. I'm always the first to say that. It's more that there was one organization and another organization and if they are not in sync, they work against one another. We were always having marketing issues with Fannie Mae because we always wanted to promote the special nature of Civano and they always wanted us to back up off of it. So we started off selling people on ideas and we forgot to sell people on the value of the home.

I give Joan Tober a lot of credit for changing the way we talked to potential home buyers. She emphasized a message of "choice and value." You get a great home; you're dealing with local builders who can offer more options to customize the house than a Pulte or KB. And by the way, here's the great news, you will save 40 percent on electricity, it has great indoor air quality, reduced use of potable water, and it's a healthier home. But this message was always placed in the context of, "Here's the value for you right now." Not, "Buy this now and be noble and save the world" but, "Buy this incredible house which is great for you, gives you value the moment you turn on the air conditioner, which is healthier for you in a neighborhood that is safer for your kids." I don't know anybody who wouldn't be interested in that. And once we started marketing Civano that way, we were competitive.

The challenges of finding a marketing message were solved largely by a professional realtor. Her views about what motivates home buyers were very grounded in practice while the ideas Civano were based on and originally marketed with were not what most people looked for in a home.

However, before a single home could have been sold, it had to be built. This means that developers have the challenge of selling their lots to builders. In some scenarios, the developer may have already installed utilities in accordance with their development plan, but the builder will have to become heavily invested in the property to get to the point they can complete sales. The builder often has to develop all above-ground infrastructure from roads to rooftops and the planted medians in between. So, how did developers sell their Civano idea to their customers, the builders?

**Rayburn:** One of the things that the development agreement required us to do was to find four builders willing to build here. And that was a very tough job. They looked at what they would be required to do and they'd say, "That is just too ambitious, and outside what we can market!" So it was a very difficult job.

After awhile they found four builders who were attracted to the Civano project for different reasons. Their reasoning ranged from one builder who was older and wanted to leave a legacy project to another who wanted to add to his portfolio already being built up on innovative projects. They each came with their core building competencies, but they needed help understanding how they could meet the energy codes without having to relearn everything they already knew how to do well.

**Rayburn:** A lot of it was about helping builders understand that it is not about reinventing the way you work. It's about being more mindful about the way you work. No, you can't go to Home Depot and buy the little Civano efficiency plug in thing and use that and be done with it. It's about lots of little moves in framing, windows, insulation, door seals—it's a whole series of things. We made a commitment to our builders that we would help them with all of this.

You know, there are those times when you work really hard and you come to a time when you think, wow, this was really worth it. So after about a year of building we had a press luncheon. One builder stood up and said, "I came to Civano with some doubt. But I have learned how to build a better house and understand my customers better." And I thought, okay, that's worth the journey. We've had an impact.

## Public Meets Private: The Civano Institute

In 1996 the state granted the newly formed Civano Institute $300,000 for conducting a builder education program. Doug Crockett, on sabbatical from the Tucson Unified School District, came onboard the Civano project full time working for the Civano Institute.

Brian McDonald's interview with Doug Crockett, 2008:

**Crockett:** Well, I'm currently the energy manager for the City of Tucson. I started that job in September of '06. And prior to that, I had been in a comparable position with Tucson Unified School District since 1991. During that time, I served on the Metropolitan Energy Commission, '91 through '98 and took a year's leave of absence or sabbatical from TUSD and worked on the Civano Institute in 1997.

**McDonald:** The Civano Institute for Sustainable Communities?

**Crockett:** Well, it was originally called the Civano Institute. And then partway through that year, we changed our name to the Tucson Institute for Sustainable Communities, partly at the request of the developer just because they didn't want that name "Civano" taken up. I think they had some other ideas for how that name could be used. I think they had in mind that they might use that name, Civano Institute, for something else that would be directly under their control.

**McDonald:** Was that a problem since the Civano Institute was not directly under the developer's control?

**Crockett:** Well, I don't mean to speak for them, but I think they had concerns about a separate non-profit with its own board of directors making decisions about what could affect the development and educational program directly.

**McDonald:** So the Tucson Institute for Sustainable Communities, what was its goal?

**Crockett:** It really was to try and promote education about the sustainable energy standard and about the performance targets at Civano and to try and broaden that awareness and application city-wide, not just to Civano development.

**McDonald:** So they were very successful in the end? The city adopted the Civano standard for municipal buildings and abandoned it to build to LEED's silver level standard?

**Crockett:** They didn't really—I'll correct you in that. They didn't really abandon it. They just basically added to that because I think the way the council worded its resolution back in April of '06, they basically built on top of the sustainable energy standard. So there's still that 5 percent of energy in future new city buildings, for example, which must be derived from solar—in addition to the minimal LEED silver standard.

**McDonald:** Okay. So you're involvement was for only a year with the Civano Institute, or TISC (Tucson Institute for Sustainable Communities) as it was later called.

**Crockett:** Well, officially I was a paid salary person for a year. I had some involvement before they actually got off the ground. And I also served on the board then for I think at least a year afterwards until they—we decided to disband to organization, I think in 1999, just because of lack of funding. So, as a member of the Metropolitan Energy Commission, we had something to do with the process leading up to the developer being selected for the Civano project. Even before I moved to Tucson in 1991, I had read about what was then called the Tucson Solar Village and the concept of a piece of property that would be kind of a showcase for solar and energy efficient design and residential and commercial buildings. And at that time, the Metropolitan Energy Commission was supporting some of the initial research. And the City of Tucson, in fact, had a full-time staff member, Will Orr, who was devoted toward developing that concept into fruition. He was pulling together the concept of a Tucson solar village and really redefining it, helping market it. So he did some of that preliminary work on behalf of the city. And then later, that was picked up, transferred some of that responsibility to John Laswick.

**McDonald**: What was the nature of the relationship between the Tucson Institute for Sustainable Communities and Case Enterprises?

**Crockett:** I don't want to talk for the developer, but I think they felt a lot of responsibility for making something happen in a very short time. And the concept of a separate non-profit with its own board of directors kind of deciding things that would be happening that would relate to the development seemed like a diversion or a problem that they didn't need to deal with in the midst of everything else. And they were, to their credit, cooperative when we actually scheduled workshops and asked for their participation; in certain ways, they would comply.

I think we had the potential—if they had been more supportive I think something like the Civano Institute would still be around today, both complementing what they're doing in Civano but also helping the entire community. And because of both the insistence on the name change as well as not wanting to see such an entity grow, it really did not—there was not financial and structural support for it to continue. And it couldn't subsist completely on grant funds, which is why then we decided to disband it.

**McDonald:** So TISC was around during a very tumultuous period of Civano's history—between '96 to '99?

**Crockett:** Well, the Civano Institute did not exist at the same time as the land auction; it came later. Officially it started I think sometime in the fall of '96 probably. And then I had direct involvement during the calendar year '97 when I was taking a leave from TUSD. And I know it continued on for at least a year after that while I served on the board. And I think we may have put the records to rest sometime in 1999.

**McDonald:** And going back to '91, you had heard about this sustainable community in Civano, and what attracted you to that?

**Crockett:** Lots of elements. I mean, having done a fair amount of work in the construction business and design business myself and the energy efficiency business, there are a lot of bad examples of how to do all those things and how to interrelate them, especially—well, look at any suburban development.

I mean, most construction, design construction going on historically has been very lowest first cost and poorly thought through in terms of the life-cycle energy cost and maintenance costs on those facilities. So it was exciting for me to see a project that was looking more comprehensively at how you could really design better, individual structures but then have them related in a neighborhood concept so that residents would be paying less over the life of those buildings for the utilities and that they would be related in such a way, the buildings would be related in such a way that people could actually be reducing their commuting times if you're mixing residential and commercial use. There could be a performance target of actually having a certain percentage of residents be low-income, they wouldn't just be a gated community and only wealthy people could afford to live there.

I think my only reservation about Civano from the beginning was that because of the availability of that kind of plot of land, it had to be quite a ways out from the downtown of Tucson. But once things unfolded and it became clear that piece of state land could be used for that purpose, then it was kind of a question of doing something versus doing nothing. And I was supportive of doing something.

And I'm still proud of the fact that Tucson as a community and the people that are involved directly did make something happen that I think was very significant to not only this area, but nationally and internationally, in demonstrating that something like this could be done.

**McDonald:** What do you think about the public/private partnership that it took to create Civano?

**Crockett:** It took that same public/private partnership to bring the Civano Institute to fruition. That's the kind of thing that with the city's support and with grant funding was able to bring something to fruition that would not have been possible without that partnership. A lot of times projects get off the ground where they are just privately funded or if they're government supported. But if you can get the best of both of those worlds and you have some private investment in encouraging those kind of things to develop plus you have some buy-in from local government to support it and provide in-kind support— those kinds of efforts can—can more often flourish.

And I really think the Civano project is an excellent example and certainly would not happen had there not been some combination of the city taking an active role in promoting it and trying to find the right developer as well as having a private developer who was going to put their money at risk and willing to take on the leadership required to implement the idea. It's one thing to have the vision. But unless someone's really willing to do the hard work of making it happen, it's, a lot of times, just going to stay an abstract idea.

**McDonald:** So during your time at TISC, did you find that there was eroding support from the city with regard to Civano?

**Crockett:** Eroding support for Civano from the city? I'm not sure I could verify that. I think my experience there was some struggling with how to make it happen. And there was kind of the challenge of how do you implement some really innovate ideas within the context of a fairly conservative kind of establishment of how permits, building permits are issued and how codes are interpreted that really I think was just an evidence of why Civano hadn't happened to date. So things were different with regard to Civano.

I think any number of issues came up that had to be resolved in a different way. And I think of just street widths, for example, and curb cuts in order to have a higher density, more walkable community. There was a lot of effort to make roads narrower and build in parking along the roadways as opposed to having really broad, wide streets that would kind of segregate traffic from bicyclists and pedestrians. And there were issues of both the energy efficiency components and water efficiency components, those kind of specific things, as well as one of the other overlays with Civano that I think made it even more interesting and, to date, keeps it a unique model is how they—we overlaid the kind of new urbanist kind of concepts and how do you make a community more livable, more walkable, more friendly, more attractive than a standard suburb, even if the buildings individually are more energy efficient or resource efficient. But how do you make it really a more pleasant place to be and a more resident-friendly area?

So to integrate all of that and have that be reflected in how the city interpreted the plans, the individual details that were going into that concept led to lots of issues that had to get resolved. And when there's a time pressure on everything because money's at stake, then it becomes, I think, part of the burden for the city was to try and fast-track some of those things or ensure that those decisions got made in a timely way so that the developer wasn't at risk for trying to bring something innovative to the table but have some support in the city to change the way they always did things so that, in fact, something innovative could happen.

Did it work? I think that yes and no. There were times when I saw compromises being made or changes being made that were a different approach for good reasons and the city was willing to try something new once they understood why it was being done and how it was going to be applied and to ensure that residents' safety or health was not being risked. And there were times when it failed just due to shared inertia of the way things have always been done and people's unwillingness to change the way they've always done it or because of the time pressure. There's—not everything got done. You know, it certainly was not a perfect process. And anybody who wanted it to be perfect on either side was usually disappointed. So there were lots of compromises and I'm sure in retrospect some bad decisions about investments or prioritization of things that I hope if at—even with mistakes being made, I think Civano can still provide a good case study for what—what we learned in the process so that people who have a similar vision or willingness to take on a project like that could learn from our experience and hopefully do it better without having to re-create the wheel.

**McDonald:** Do you know anything about the problems with the learning curve in Civano?

**Crockett:** Only peripherally. I mean, part of our charge was, you saw with this kind of building program, was to educate builders as well as residents or owners in terms of what the—especially what the performance targets were and how to achieve those. Because I think at the time, even though in retrospect it seems like it's pretty straightforward, or now that we're doing it as a matter of course, it's not rocket science anymore. But at the time, people, I think, and builders especially were afraid of what this meant and were at the assumption that it meant that they had to spend more cost per square foot basis. And even though people talked about energy savings and water savings, I think there was general skepticism that it was worth the additional cost. And there was an assumption that it had to cost a lot more in order to show those savings. And what we tried to communicate in that—in our education program and I think has persisted now as you look at the LEED system and how—how that quantifies what a green building is or a resource efficient building is and how to achieve it mostly, is that a lot of the secret is knowing where you're trying to get to and starting that early enough so that you could integrate good design ideas together with the technologist to make that happen. Whereas I think at the time Civano was starting, people were still kind of looking at green buildings as something as an add-alternate almost. You built a conventional house and then you added on some technologies or some secrets that would somehow make it more energy efficient or water efficient.

That can be successful if you have an unlimited budget or if you just look at buying your way into efficiency. But the secret really is, as I say, having a vision of where you want to be and where you want to get to and then integrating a lot of the design elements so that the client gets what they want and gets a livable building but that you do that through a number of ways that any one solution can solve a number of different things all at the same time. And being able to incorporate those various solutions into something that can really not—does not have to cost more money but has to be more successful. It has to solve the clients' needs without spending a lot more money. And I think the builders that responded to that or saw the opportunity there and started solving those kind of problems and realized how to participate in achieving a better design with less cost are the ones that were the most successful. Whereas the builders that wanted to keep doing things the way they'd always been doing them but wanted to get credit for being a green builder at the same time usually struggled and may have ended up not continuing just because it was too much effort or they never really got it, what was different about Civano and about a much more integrated design process.

**McDonald:** How would you define a plan book?

**Crockett:** You know, I should have given you an example of that, because we did actually publish a plan book. And what we try to do is take I think five designs from Civano and almost treat them as case studies for how—how the design incorporated the performance targets, especially how it achieved that. We also took five designs from existing Tucson homes and again showed the same elements but outside Civano. And that was really, I think it's one of the best things the Civano Institute did—or by that time it was called Tucson Institute for Sustainable Communities. And we wanted to make the point that individual residents in Tucson didn't have to be in—living in Civano in order to achieve the same performance and that there were good examples of that already existing. And there was every reason for people to think about the potential of doing that in their individual properties, or there's certainly reason for developers to think that with in-fill projects or bigger scale developments.

So I think to me that was a useful—useful document and I think there's currently talk about green building kind of being revived through the Southern Arizona Home Builders Association through Pima community—or Pima County. But there have not been—there have not been any real active steps taken to codify that or to certify that in a way that I think that we desperately need in this community because there's still—the norm is still building lowest-cost housing at some level. And that's the only measure of the value to the owner. And there's really not a way of quantifying the value to the community if you look at more of a life-cycle cost approach and even have both incentives to achieve that as well as disincentives to discourage the really bad design and construction that's going on that becomes a drain on the community.

**McDonald:** Do you think it would have been easier for everybody in Civano had there been a plan book from the very beginning, from the land auction?

**Crockett:** Very likely. Sure. I mean, having some—having a blueprint in place is always easier than starting from scratch. And that's a woulda, shoulda, coulda kind of statement. And it's always nicer if you had something. But to their credit, I think those people that weighed in, including the builders, the developer, the city, there was I think a healthy level of creative thinking going on and struggling with the problem to come up with solutions that could work in Tucson, Arizona, to achieve these performance targets. And to me that's one of the biggest successes of Civano, that people weighed in I mean, people in the community came up with a vision for how it could be different; and other people responded and made it happen on some level. And it was not—I don't think anybody presumes that it was a perfect solution. But it is, I think, an example of something unique that happened in Tucson and that could be expanded upon, it could be perfected with other iterations. So I have no regrets about what it did or even how it was implemented. And I think a lot of the people that put time and effort into making it happen deserve a lot of credit for it, even though it wasn't perfect.

**McDonald:** Do you view Civano as a success?

**Crockett:** I think it's a success. Just the fact that it's—it happened and it's still a functioning community. I think it would have been a failure had we aborted at some level, had we stopped midstream and thrown up our hands and said we can't do this or nothing was built or that it was built not to performance standards. I mean, any number of things could have had a tip over into the failure side. But as far as I'm concerned, it was a success for what it was, how much support it had. And I hope your interviews can help us learn more about what made it successful and how something could be made more successful based on the learning from that experience.

# Lee Rayburn Summation

**Rayburn:** Look, when you take away some of the difficulties of Civano—its early project R&D aspect; not knowing exactly what we were doing because nobody had done it before; some management decisions that were, in hindsight, questionable; and others that were forced upon us—this project is not some huge raging torrent of red ink. We had some problems. We probably overdid some things in the design. It's important to note that this project has sold well and its homes have retained great value. If you bought a KE&G house in first-phase Civano you could get a house for $179,000. Second phase KE&G homes went up to about $220,000 and we all thought that seemed high; and yet they sold! That was one of the times when we realized that the look and feel of new urbanism combined with energy savings was very interesting.

One of the lessons learned in Civano is that a lot can get lost in translation. One example was that the urban design of Civano included a heavier natural landscaping routine inside center medians and as buffers to the sidewalk areas that were, through the public improvement agreement (PIA) process, to become the property and responsibility of the City of Tucson. Upon viewing the plans, the City of Tucson realized they did not want to take on the added cost of maintaining these heavily landscaped medians—yet the language was never struck. So when a tree root heaves a sidewalk, nobody is quite sure who pays for the repair. So far, the city has paid the bill.

This is a classic example of the grey area between code and pattern language. The pattern language suggested it would be good to design with more natural, vegetated areas that can serve multiple purposes such as drainage areas, shade for the streets and pedestrians, and safety buffers from vehicles. Unfortunately, there was no code language to determine what constitutes the city's upper limit on maintenance costs for a right of way. In later stages of Civano's development, such as Pulte's Sierra Morado, the issue of right-of-way maintenance was clearly defined. The HOA pays the bill.

# 6

# TUG OF WAR

## *Rediscovery*

> *A man builds a fine house; and now he has a master, and a task for life; he is to furnish, watch, show it, and keep it in repair the rest of his days.*
>
> —EMERSON, "Works and Days," *Society and Solitude* (1870).

Why is it that people often forget suffering? We obviously don't forget the experience of things not going our way, or being hungry, being injured, or going broke; but pain itself usually goes away. We don't remember the experience of our fear as a vivid thing, something tangible-only that we were afraid. Especially when injury occurred over a generation ago, as a society we often let it go into a fuzzy-memory place where the meaning of the event itself is washed and faded. Thus are the pains of our fluctuating economy and the resulting innovations, so it seems. Good times behind us often mean harder times ahead for Americans. The Civano story paints a picture of the affects of boom and bust economy on the building industry, and on building innovation. One of the City of Tucson's project managers for Civano, John Laswick, illustrates this point nicely.

**Laswick:** By October '91, they brought Civano to the city council for zoning approval and approval of the master plan ... and Civano didn't have a negative image like the mainstream sprawl going in just down the road. You know, everybody loved it; people were all behind it. And so in October the city council approved it unanimously; and meanwhile, the real estate market was in the worst shape it had seen in 20 years. So builders were going out of business right and left, commercial developers were filing for bankruptcy.

Tucson actually lost population one year. At least, we were barely gaining population. Normally, we've gained about 20,000 people a year. So when it drops down to about 2000 that's pretty serious. You know, it was a national slow down going on. This was like the tail end of all the real estate zaniness of the mid to late '80s when people were just building anything right and left because of the tax laws. They made a lot of money just to build anything, even they didn't have tenants.

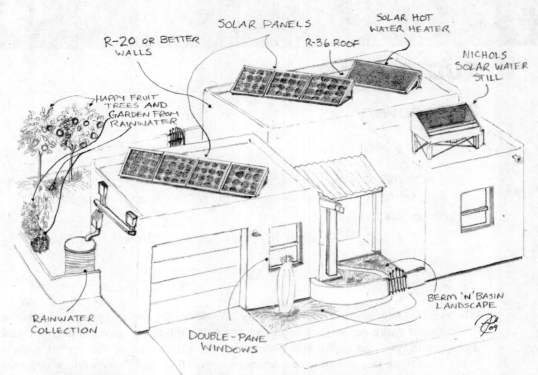

**Figure 6.1** A home in Civano with solar amenities, rainwater collection, and high performance building envelope.

So Tucson had thousands, tens of thousands of square feet of vacant industrial space and retail space. IBM was leaving town. Well, they never quite left, but they went from 5000 people down to something closer to 1500. And the housing market was bad. So the bottom line is nobody was doing anything as far as new development was concerned. In particular, not something like Civano, which was actually costing more up front because it was different and had more amenities.

**McDonald:** A solar hot water heater is an amenity?

**Laswick:** Yeah. Individually those things don't cost that much more; but collectively they cost somewhat more. I realize that they save the homeowner money in the long run; so I think that is the tradeoff: a little bit more laid out up front for something better in the long run. And I think that can be sold, as well.

But when you talked to builders then, they were not really interested in doing things that were different. You look at the innovation curve in computers, it's something like 18 months between technology jumps—from cycle to cycle. It took something more like 15 years for the building trades to absorb steel framing technology because so many builders were hammer-and-nail guys—that is what they knew. Builders don't want to screw around with stuff that they don't understand because it is an unknown, a potential risk.

So things happen very slowly in real estate, they happen, but they're often very expensive. Basically what you're talking about with Civano was something that very few people had ever seen before. I mean, a few people had been to Village Homes, but that's a subdivision scale. That's 70 houses I think.

**McDonald:** And far outside the price range of affordability.

**Laswick:** Not really, not then, no. It's more expensive now; but it wasn't when they built it. Neither was Civano, especially initially. But anyway, the problem is that you have a project that's in the $50 to $75-million range, probably closer to 75 as I recall it. And a developer would look at that and say well, that's really like a $65-million project, and you need 75—so you're $10 million short. So what do you do? In order to induce the innovation, you have to prime the pump with some public investment.

And so Wayne's contract was running out at the time and the money was kind of running out, so they went out looking for foundations or other organizations to put up some funding. But nobody was giving $10 or $15 million. They just don't give you that kind of money. And the other thing, of course, was the backdrop of the '80s. There was all this Carter-era stuff leading up to big investments into solar. And as soon as Regan came in, he said "forget this" and took the solar water heaters off the White House. So you had about a decade of growing interest and support for solar and then it hit this brick wall. That lasted about until Clinton came along. So, Carter had gotten some tax breaks passed; and then people were able to afford to do some solar and efficiency stuff. Those got abused, which was one of the reasons they said they were going to shut them down.

Some contractors were just going around installing solar stuff willy-nilly. I've got a friend that works for Southwest Gas that told me he went up to this rooftop one time and saw that the solar collector, the hot water collector, was installed upside-down. Things like that. So people were just kind of doing it for the tax breaks. These programs discounted the price so much that it was ... anyway, that doesn't matter.

Civano was an idea born in the 1980s and planted in the 1990s, but it was still being built through the writing of this book. There was the dot-com bust of the 90s and there has been other economic waxing and waning such as the recession of 1983 and "Black Monday" in 1987. But for much of its time thus far, Civano grew during a good climate for building. For years our war machine was not necessarily idle, but it was not drawing down reserves at the frantic pace of the first decade in the year 2000. Concurrently, housing prices were going up and loan requirements were falling down. It was progressively easier for previously non-potential home buyers to become home owners (or at least mortgage-payers).

Then the Twin Towers fell, we watched parts of New Orleans and many other places in the coastal South get diminished to rubble, our troops got longer tours of duty in Iraq and Afghanistan—and then the banking collapse. At the same time we were challenged by the juxtaposition of a widening environmental awareness. Once again, fossil energy prices ran upwards and the quality of life for many people across the globe worsened. Al Gore's "The Inconvenient Truth" took on the established paradigm and helped us consider that we have become a singular environmental determinant, much like Carolyn Merchant had done about four decades earlier.

Our oceans are rising and will rise further. Our glaciers are melting. Our storms are larger. Our machines and our bellies are extremely consumptive. And yet, our perspective is somewhat short. Although we have developed excellent methods of extrapolating information about the world beyond what we have seen and documented first hand, the oldest written words and purposeful records to be found from people on earth are but a few thousand years old. Paleoclimatology research helps us look back much further and recognize that the climate has been changing and often drastically. Thus, we need to know how are we going to deal with it as it happens because there is little question *if* climate will change in ways that challenge our ways of life. No big deal, just business as usual on a planet that constantly changes, with or without us.

Since long before the conception of Civano, people have been gathering pace rising to this new perspective and accepting that no matter if it is our fault or not, we need to understand why our environment is how it is and we must adapt to become the best environmental stewards we can be. We may need to be able to manipulate the climate expertly someday to counter disastrous natural effects. Perhaps it may even become necessary to release greenhouse gasses rather than sequester them, or is that only science fiction?

At the time of this writing, times are strangely like they were in the early 90s as described by John Laswick—only worse, and better. Building in Civano has slowed only somewhat less than within Tucson at large. The economy is depressed and the price of gas has gone down to below $2.00/gallon in Tucson. Tucson motorists are speeding up again in their half-value pickups and SUVs. Nationwide unemployment is over 8 percent and rising daily. The electric bicycle shop proprietor in Tucson is living in fear that his shop was opened prematurely, that he misjudged the great green wave of change.

At this same time, a new and progressive president is taking the helm. What will it mean for Civano and the green movement? The answer is in our hands.

President Obama was elected to the highest U.S. office sharing a vision, a new ideology. Even when the economy is seriously struggling, he is promoting sustainability and the greening of America's infrastructure, saying we cannot afford to do otherwise. This may be a complete departure from the normal order of things, and changes the place of green in a bust and boom economy from being expendable, to one of the top priorities. President Obama's actions made it clear that the means is also the end. So we cannot destroy the means solely to reach an end. The process of what we do is part and parcel to the reason for doing it. We develop and participate in the functioning of society because that is how we provide food, health, shelter, and a modern lifestyle to each other, and ourselves.

This kind of integrated reasoning about the function of America's infrastructure parallels the logic behind a program like the LEED certification process. The United States Green Building Council is behind the vision that "buildings and communities will regenerate and sustain the health and vitality of all life within a generation."

That is an incredible goal, yet nothing short of what needs to be done. If America buys into this option for change and does not sell out early, green building will flourish in the United States and Civano will one day become beautifully mute—a fine and serviceable old-growth pioneer to the cause and a fully-cleared stepping stone to change. Green will once again denote a wavelength of visible light, roughly 512 nanometers or so—and a heritage of saving life's diversity on Earth.

# Choosing a Master Developer

**Laswick:** And so that's about where I came in—late '94. The Metropolitan Energy Commission at the time was advocating that the city allow them to search for a master developer and to conduct a search for that particular somebody who would actually do this job. There was about $650,000 left out of the some million-sixty originally granted to Civano planning at that point, which was pretty good. And you know, there were some different thoughts about how that was supposed to be spent. One idea was for a demonstration project, which actually ended up being the community center in Civano.

The city council decided that they didn't think the Metropolitan Energy Commission was the right organization to try to figure out who should be the developer. It wasn't a hit on the MEC because they knew about energy, but they didn't know about real estate development. So the city council directed the city manager at the time, a guy named Mike Brown, to essentially get this thing done, you know, to find a way to make this thing work. So then I made the mistake of walking into Mike's office the following day saying that I was bored. And that's how I got involved with the project. I had a background including real estate development and real estate finance. And I was working for the city's economic development office; and I considered myself an environmentalist. I lived in a passive solar house at the time, so I thought I was about as well qualified to do this.

The Metropolitan Energy Commission, really would have preferred to have done it all themselves, but like I say…

**McDonald:** Why is that?

**Laswick:** Well, they thought it was their baby, and it was. It's just that at some point, your baby grows up; and it then it needs something different. They had nurtured it; and they protected it; and they grew it; and it got to a certain point, and it left home. I think that is when Civano needed a different guide. But, I go out of my way to give the Metropolitan Energy Commission members credit.

You know, now that I don't work for the city, I can see that the city does sometimes tend to want to run everything; and I don't necessarily agree with that tendency. But on the other hand, you could say it wasn't MEC's project. The city had been kind of sheltering the project for the previous couple of years. And at that point, some of the city council members just decided MEC had done a great job to that point; but it was time that Civano needed something different. And so the city decided to kind of take it on and move it to the next level using its knowledge and expertise in real estate and land development which wasn't necessarily a part of MEC's core skill set.

After those initial stages, it becomes a different process. The city is used to running requests for proposals and doing this kind of work. They also understand building codes, energy requirements; and they understand infrastructure development. I don't think anybody would argue that the city should have been the developer and developed the land itself. I don't think anybody would have said that. I was playing this kind of intermediary role between the state, which has the money but doesn't really have direct involvement with the project; and the Metropolitan Energy Commission, which had the vision but didn't really have the real estate background. We were essentially a transitional agency.

So, we were putting the pieces together, trying to get enough information for a real estate developer to look at the plan and convince them it just might work. I assumed going in that 80 percent of them wouldn't even bother wasting their time thinking about it. But you know, all we needed was one.

Then about the same time that was going on, we started promoting green building and trying to see how many builders in town would actually go for this. Civano was a master development project—and the city wasn't soliciting builders. The city was trying to get somebody to do the infrastructure—the streets and the sewer line, all that stuff; and then the developers, in turn, sell parcels or blocks of land to builders. And so we were already, in a way, one step removed.

I knew that anybody who might be looking at this, particularly if they were not from Tucson, would wonder who was going to build these houses. Because you could say what you want; but does anybody know what our goals would really look like in an end product? So we brought some people in from Austin, from their green building program; and we did a presentation. And with that effort, we collectively had a more complete overview of the ballgame. Well, we knew where the outfield was, but we didn't know where home plate was. So you know, it was all starting to make some sense, but not completely, yet.

Also around that time, it was actually early in '95, I was calling around the country to see who else was doing this type of project. The good news, and the bad news, was that not many people had experience with this kind of effort. So I called one person who I knew had been involved with some kinds of forward thinking development ideas, David Butterfield. He said he'd heard about this development, that it sounded really interesting, and it sounded very much like what they were interested in doing.

So he came down, and we spent the day driving all over the place. I introduced him to a bunch of real estate professionals and local builders—and he really liked it. And you know, even though I knew about the new urbanism, I hadn't connected it with Civano in the way that David had already thought through. Because he really believes in the environmentally friendly approach; but he also believes in the design and the human scale and the human feeling of new urbanism.

And a light went off for me, like, "Oh, of course, why didn't I see this?" And you know, to me new urbanism is the curb appeal part of it because with any luck you don't even see those darn solar hot water heaters. People on the MEC would like to look at those solar hot water heaters. But most people don't really want to see that.

And so it was just at the time when new urbanism was really kind of coming along. Well, there were still very few developments that were being done, but it was getting a lot more talk and more public awareness and so forth. To me it was a hand-in-glove sort of situation because new urbanism communities are intended to be very walkable. They have shade. They have all these built in amenities and in fact, after Duany got involved with this for a while, I sat down in a room with him; and I said, "Andres, I just don't understand why new urbanism doesn't claim more credit for being environmentally friendly because it already has that going for it." Why wouldn't they want to have more marketing benefits? You know, and his answer was, "Oh, well I think the environment's in good hands." Well, hmmmm…

He's a brilliant guy; and it seemed to me that for him, design trumps everything. So what he does is beautiful design; and he does it really, really well. But, there is this whole other realm of new urbanism that fit into the Civano goals and scenario just so perfectly. So it was at that point that we really made the connection; and it was really David that kind of brought that into it.

There were people who didn't like that. They probably still don't like that. They thought that the new urbanism was kind of a kitschy sort of fad. To them, Civano was supposed to be about energy efficiency first and foremost. There are many people who focus on the function of a place like that. And they're great people; and they're totally dedicated; but they're not mainstream homebuyers. I don't think any of the original Metropolitan Energy Commission people live in Civano, for example. Well, with the exception of Al Nichols, that's true. But he's the only one probably.

Interestingly, extremely few of the people involved with the early development of Civano ever ended up living there. Perhaps this goes back to Lee Rayburn's lesson learned about sending a consistent message about what the project is and what it intends to do. To the date that Civano was seeking a master developer, there had been some new urbanist master planned developments, but none that reached for the sustainability goals and energy standards derived for Civano. As John Laswick points out, the beautiful designs that are often yielded from new urbanist architects are a synthesis of form and function that happen to lend well to many environmental concerns as well. But to many people, the message was lost, or at least diluted.

The image of a couple undertaking a home remodeling project poses an interesting comparison to what happened with the Civano decision making process. Imagine the couple has a strict budget to update their older home. He is interested in creature comforts, style, and an open floor plan while she wants to save the environment by making their home carbon-neutral. The confining budget leads to an ideological tug of war about which amenities have the highest value. Both of their approaches improve the quality of life in their home, but the budget can simply not pay for every aspect of each of their wishes. So, in a fair a just world, he gets about half of his style, and she gets about half of her performance.

It is fair to say that this is a bit like what happened at Civano, except there were far more negotiators at the table, and the tug of war probably didn't end up in a 50/50 draw. So when the project was completed it did not fully resemble very many of the participants' conception or vision of full success, leading to frustration on the part of many. However, the homes did sell well to individuals much less involved with the project conception, planning, and design.

# Market Study for Civano

**Laswick:** And so at that point I worked out a plan to do a market study because nobody had done a market analysis of this thing.

**McDonald:** How do you do that?

**Laswick:** You start with a request for proposals (RFP). So basically, we just advertised for the help we needed. What we advertised for was an analysis, but not a conventional real estate type of analysis. We already had one of those. What we realized was that any real estate developer would want to know more about who was going to buy the Civano homes and how much are they willing to pay. Every competitive real estate developer does market studies before they proceed.

So, first of all, we got a market analysis [sourced out of] Phoenix; and then they had a subcontractor that actually did interviews. They did random samples of 300 households in the Tucson area. They called people up. And we told them what some of the questions were that we wanted them to ask. We wanted them to ask if prospective home buyers would be interested in a home that had these specific kinds of features: like solar, increased envelope efficiency, etc. We also wanted to ask about what community amenities they wanted, and what kinds of features were most desirable. Like open space, walking trails, a community center, etc. Actually, the market research firm was out of Scottsdale.

At first they implied nobody was going to want to buy this stuff… and then it came back that something like about 60 percent of the people said, "Yeah, we'd like that." They even said they'd be willing to pay $5,000 more for a house that had those kinds of features. And so then the challenge became: can it be built for only $5,000 more? But it became clear that there really was willingness to trade off higher front-end costs for better long-term performance. We didn't steer them in that direction. I mean, I think it's pretty logical, people get it. I think when it is presented the right way, it gets to be pretty instinctive.

**McDonald:** I'm sorry, I just want to clarify one thing. You said that it's pretty instinctive, like to use the sun instead of burning coal?

**Laswick:** Correct. Or to save water in the desert. People in Paradise Valley might not be in an environment that pushes them to understand that, but people in Tucson do. And so at that point we had much more information. Then the other thing that we started to do was to promote green building.

Petersen, Jacuzzi & Green provided a feasibility study of the Civano project and provided specific market-driven suggestions for completing the successful development of Civano. This report shows some interesting trends regarding how people viewed green housing in the late 90s. The average recent Tucson homebuyer at the time, according to PJ&G's report had lived in their home for nine months, mostly located in the northern and central portions of Tucson. This group had sold their previous homes for an average of $132,200 and made their new home purchase decisions in approximately 4.5 months. Their homes mostly ranged from three to four bedrooms, two baths and just over 2000 square feet of living space. These individuals paid approximately $184,341 on average for their new home, likely to have amenities such as swimming, golf, daycare, and security services.

The most important community features to these recent home buyers were the investment value of the home, views, type and price of the home. Desirous community improvements were reported to be a community center building, a gated entry, and a trail system. This group reported high importance and value on the home itself, peace and quiet and quality of lifestyle. Security, maintenance of open spaces, and the distance to parks and shopping were lower on this list.

The emphasis put on features such as walking trails, quality of lifestyle, and distance to shopping starts sounding a little bit new urbanist in nature. These are some of the things that may be more instinctual; and why they work well as design features. However, these same home buyers who reported high value for open spaces, walking and jogging pathways, parks, and proximity to shopping also valued larger lots, single family dwellings, and the demand for an extra room: the home office. Privacy was a strong desire, equating to the desire to have larger distances between dwellings—even if it comes at a cost. This group stayed away from condominiums.

Interestingly, these individuals reported that they liked the idea of having mixed housing types, even though they would not choose to live in multi-family dwellings themselves. They want pools, but for the most part would pass if they had access to a community pool. The recent homebuyer survey responses did not reflect a specific concern for the environment, rather an amenity-driven decision making process. After surveys of several cohorts, PJ&G's report suggested that Civano's potential for market success was very high. The fact that Civano was to have trails, parks, and local amenities was most certainly some of the attraction. However, so many of Civano's homes are in some ways exactly opposite of the image portrayed by the recent homebuyers in PJ&G's study.

The respondents wanted larger homes and larger yards to buffer them from their neighbors. But Civano, in its quest to be green, reduced most of the home footprints, shrank down the yards (in some cases to elimination) and there are areas where privacy is not really an option, with a neighbor's balcony poised with a bird's-eye view of at least one or two of their neighbors' courtyards. Civano was to be a much more urban-feeling housing development whereas the homeowners surveyed in the late 90s were visioning suburbs with larger lots and amenities as their ideal places to live (given their income).

How do these late 90's trends compare to the most recent homes sold in Sierra Morado, Pulte Home's development in Civano? Excerpts from Gal Witmer's IMPACT System Monitoring Report #13 help tell that story.

Sierra Morado Pulte Homes completed a total of 126 homes in 2007 bringing the new total in Sierra Morado to 341. The average sales price in 2007 was approximately $245,000. Pulte continued to offer the 2400 series in the Las Colinas series—the La Estrella model, 1356 sf, 10 sold with starting price of $162,190 and the Alta Mira model, 1479 sf, 24 sold with a starting price of $172,190. These models, the smallest of all offered, are located on their lots similarly to homes in neighborhood edge in Civano 1 with garages accessible in the rear via alleys and landscaped pedestrian way front yards. The smaller size of the Alta Mira model combined with its affordable price and three bedrooms made it one of the primary contributors to the affordable housing requirements in 2007.

Pulte offered a number of new models in 2007. The El Paseo series offers four new models: The Andalucia, 1448 sf, $188,990 starting price with nine sold; the Seville model, 1653 sf, $198,990 starting price with 10 sold; the Trevelez, 1769 sf, $216,990 starting price with five sold; and the Portugas, 1870 sf, $223,990 starting price with five sold. In the La Mesa series two new models were offered. The Verde, 3549 sf, a starting price of $372,990 with 12 sold and the Pecos, 4071 sf, a starting price of $407,990 with 22 sold. The final new series offered, the Tierra Calida, had three new models. The Sahara is 1517 sf, has a starting price of $235,990 and seven sold; the Mojave, 1880 sf, $248,990 starting price and 15 sold, and finally the Sonora, 2188 sf, a starting price of $260,990 and seven sold.

The Sierra Morado Community Center was completed in 2007 with a grand opening in July. More than 1700 people attended the opening which featured a live concert by the Gin Blossoms. The center features a pool, tot lot, basketball courts, meeting rooms, demonstration garden, event lawn, and promenade.

Pulte provides a very large range of home sizes and prices in Sierra Morado. The sales numbers are interesting in that the smallest home model sold the most, at 24 units; whereas the largest model was the next most sold, at 22. The next closest number of sales came in at 15 and was for a middle-sized home. Does this suggest that there is a split in the trend of recent home buyers since the research done by PJ&G? Although this comparison is short of a scientific study, it appears that in Sierra Morado, there is as much of a market for small homes as there is for larger ones and that the draw to amenities is still good. But Pulte does little to advertise anything about being in a green community, placing equal or more focus on other ways of addressing the strengths of their housing products.

## Why Fear the "S" Word?

Kevin Kelly, at the time of this interview with Brian McDonald in 2007, had already planned over 8500 acres of land and completed over $3 billion in development. Portions of his story about Civano detail the complex developer, builder, and government interactions that can challenge innovation—as well as successes to be realized. Ironically, a decade later sustainability is an honorable word and something consumers increasingly respect and support. Mr. Kelly has taken the seeds of Civano and is planting them far and wide.[1]

**Kelly:** I was the managing member of Case Enterprises, which later became the Civano Development Company. So, I was the managing member of the initial development company that took Civano from a policy concept into reality.

**McDonald:** What was the path from concept to reality starting from the time of the land auction?

**Kelly:** The city had an RFP (request for proposal) prior to that land auction and John Jones was really innovative and key to this whole thing. Part of the RFP said that anyone that would meet certain design and energy code criterion would have a development agreement with the city of Tucson. If you have a development agreement with the City of Tucson, you have an advantage, functionally, going to the state at the public auction to bid for the land. It was John Jones who instituted this idea of linking the development agreement with the city which brought with it certain financing incentives to the private sector so that they could collectively use that as leverage to go get the land. It was that leverage John came up with as part of the city's RFP that made sure the most likely successful bidder on the state land would have already been committed in advance to develop a sustainable community. And David Butterfield, to his credit, worked with John to create that RFP prior to the bid process.

David Butterfield with his company wanted to be a bidder in addition to being a consultant on the RFP because David needed more strength, economically. Therefore, we partnered with David Butterfield together to go to the auction and win the state land bid.

**McDonald:** Do you recall any of the specific incentives associated with the land development agreement?

**Kelly:** The specifics associated with the Memorandum of Understanding with the City of Tucson brought with it access to certain public infrastructure financing,

---
[1] Visit Kevin Kelly's website: www.civanoliving.com to learn more about his efforts.

accelerated entitlement approvals, the establishment of a city ward district office in our mixed-use neighborhood center, some off-site improvement money, and a belief that we were in a partnership with the city. There was also the anticipation of some state solar funds for our development. Please understand that the public funding also brought with it additional government monitoring and project involvement. And while the intentions were honorable by all parties, the city slowly withdrew much of their support during the life of the development. So we bought the land with the development agreement which said we were committed in advance basically to build the first sustainable community in the United States.

**McDonald:** Did it work?

**Kelly:** I can only tell you the facts. I cannot tell you the myths. The facts are that it worked extremely well in terms of its acceptability in the market place until there was a basic conflict between the management, ownership, and government around the vision. I can tell you that the initial receptivity, and all indicators are that that concept could have easily accelerated throughout the region. The basic barometers one would use to measure market acceptability were strong; such as buyer traffic to our information center (5000 people on opening weekend and 20,000 visitors during the first few months); minimal price resistance—even though our costs were higher on a square foot basis, our energy envelopes the operating costs were much lower than the average house; the number of upgrades people purchased with their new homes indicated a willingness to spend; the interest in the energy saving technology and a general sense of shared values or community among the current homeowners was encouraging.

**McDonald:** What prevented that?

**Kelly:** I think a lot of things—and just to understand, you're talking about 5000 people walked through the community center the first weekend and 20,000 within the first four months. It was a huge success. But I think it was really hard to be the first one out of the blocks to realize this potential because everything about it was new and everything about it was a bit of a struggle. Sustainability was really the culmination of a lot of individual or collective group vision. So, it was about solar-capital, water harvesting, it was about land preservation, it was about building design, it was about land planning and pedestrian interaction, it was about mixed-use development, it was about employment generation, it was about upgrading the aesthetics of the tract house.

**McDonald:** What about affordability?

**Kelly:** Well, it was very affordable in this market place.

**McDonald:** Was the grading of phase I a large cost increase?

**Kelly:** No, how did that math work? You're asking the right question, but the supposition is probably not correct. Meaning, if I live on a 3 acre lot, or even a 1.5 acre lot it is better to not mass-grade because I can find within that 1.5 or

3 acre lot enough building footprint and work within the contours of that site without disrupting the integrative ecosystems of that property. So, therefore, I agree with the premise of your initial question. Then, you get into the idea that it is a better way to do it, purely in terms of not disrupting an ecological system. There is no debate on that, in my opinion. That's one objective. Then there is a different objective. If everybody in Tucson lives in a 3 acre lot, you'd actually have to put the people who currently live within the footprint of Tucson on a footprint closer to the size of the county. There is an interesting tradeoff here. Is a 3.3 acre lot better for the environment or is it better to closely cluster the lots and leave large tracts of undisturbed land? You just extended your roadways which disturbs ecosystems. You just broke up the flow and wild habitat migration corridors. Sustainable planning is always a question of finding the balance between the social, economic, and environmental impact of your plan. Density, done correctly is probably the most efficient way to build society, provided that by creating the efficiency of density it also allows you to have a beautiful built environment, because you're pooling common area/public area resources and leaving significant open space on the perimeter. Lower impact density, such as the 3.5-acre sustainable home that I live in, treads lighter on the land but costs more in infrastructure and is a harder layout to build a sense of community. And not everyone can afford the luxury or cost of so much open space. It's all a balance.

The challenge of land planning is deciding how much land you consume and how far you extend your infrastructure. By the way, there is validity on both sides of the argument. For instance, if I wanted to build affordable housing, then it does more for me to have the housing in a cluster design, but leaving lots of natural space around it for native flora and fauna. If I want to have the least undisturbed view corridors for rich people, then I develop a bunch of 3 or 4-acre lots where I can spot million-dollar homes throughout the hillside. I'd have a much less organic road structure and I'd be eating up the land, extending the roads, sewer and power lines and so forth, but I have a nicer view because I'm not clustering the housing. They both have value, but on the affordability side, I'd rather cluster housing together but leave the remaining land undisturbed. It depends on what the objective is. Here is an example: Is Tucson helped or hurt by not having enough density downtown?

**McDonald:** Downtowns are all hurt by not having enough density.

**Kelly:** Okay, so by allowing the low density development of Tucson over the last 35 years, we've actually spread the population out further but have essentially mitigated our opportunities to have centralized transportation corridors, mass transportation, and higher density downtowns while preserving much larger open-space tracts around the city. Still, it depends on what the objectives are. I tend to agree with what you said. There needs to be more density downtown. And, by the way, it should be more interesting density than tract housing design. But, from the land planner's point of view, we always come back to the objective. For mass transit, for centralized education and employment, and for fixing land values, a land planner should have a density code. It should have urban, suburban, rural, and sub-rural zones. Inside those zones, you should define your densities and land planning accordingly.

**McDonald:** Can you tell me about the interplay between the developers and builders with regard to Civano?

**Kelly:** Sure, let's start with talking about the difference between conventional builders and alternative builders, because they are actually different. The challenge for the conventional builders was to build a house that was going to be 50 percent more efficient than standard housing. A guy by the name of Al Nichols should get much of the credit for spearheading the modifications they did to the Model Energy Code. There were others involved but he really made that happen. So, the traditional production builders were a little uncomfortable to a lot uncomfortable with being innovative around an area of energy efficiency and new urbanism design. Because to get a building envelope to 20 percent above the Model Energy Code, is relatively easy to do. But to get it to that 50 percent, you had to think about it differently, have a different envelope structurally, and a different building pad. The challenge with production housing is the average builder lives on a 20 percent gross and 10 percent net profit margin, if they're lucky (back then, not today). And that's only after they build enough repetitive homes that the cost efficiencies can be realized. Asking them to innovate, change their models and infrastructure design and the manner in which they underwrite and finance their developments is asking them to take a lot of risk, initially—even though we believe changing the suburban sprawl and energy patterns will ultimately benefit society and be economically feasible. The public sector has to help drive this innovation and reduce the initial risk.

The alternative builders, on the other hand, operate in the craftsman-era model. They build a few custom homes with unique materials without the benefit of production cost savings. So, it's hard for them to shift from one-off home building to a production system that requires more initial cash outlay until they have enough volume, repetition, and production infrastructure to enjoy some cost savings. So even though this group of builders is quite innovative, it was hard for them to move from custom homes to semi-production building, and even compromise some of their craft to achieve certain cost savings. The true story of Civano is that everyone had to modify their practices and viewpoint in order for us to make this important cultural paradigm shift.

**McDonald:** So then we see the engineered buildings of Civano.

**Kelly:** Our strategy was to shift the risk burden associated with innovation over to our side in order to bring builders into the development. We did this by paying for the sustainable designs of the new home plans. We said, "We'll work with architectural and engineering experts in the community and we will design the prototype home and we will give you those plans." So we were going to take the risk out of the innovation. That was the first step with the builders. That was successful at the time. What we found was that the medium sized local builders were willing to give it a shot. These guys were builders that only built 125 or so homes, max. These were all smaller builders. They were intrigued and a little nervous.

Then the second step with the builders was to address the cost to do this. It was projected to cost about 15 percent more per square foot. They were wondering if people would pay. Eight percent of that cost was associated with energy and water savings, and 7 percent of that was associated with new urbanism and design. You know, detached garages, front porches, that sort of thing—just better aesthetics. So the builder was concerned that in a price-sensitive market that people wouldn't pay 15 percent more for the same house.

Our market research said that they would and what we found in phase I, not only did they buy at 15 percent more, but there was an average upgrade of about $30,000 per house. So the good news is, the market responded very positively.

The third issue was about active solar. The builders were afraid of active solar. We said to them that the first phase did not require active solar. But you had to plumb and pre-wire for active solar. So you had to pull wire for PV and you had to plumb for solar hot water. What happened was, within the first phase, most people wanted the solar hot water. So most everybody had solar hot water. Then we said to the builders that since the market accepted solar pre-heat water as an option, we're now going to make it a requirement. So we backed them into getting over that hump. I actually think, to the builders' credit, I think a lot of them had fun with it and responded well within the reality of their own business world.

The mid size builders were competing with the mass size builders, like KB homes, US Homes and Pulte and they didn't have the buying power, so they had thinner profit margins to begin with. So, that was okay.

The fourth factor is that they were a little uncomfortable with centralized marketing and a centralized sales center. But we convinced them to do a centralized sales center but we capitulated on having their own sales staff selling their own models.

Another point of integration and planning for a master planned community is the way it will be sold. The project was originally under one name: Civano. The different builders there had their own models and branding, but there were not differently named sub-communities such as Pulte Home's Sierra Morado. The centralizing of marketing was threatening to the builders because they spend a lot of their energy developing their reputation, name, and overall brand recognition. The idea of all companies under one marketing roof, figuratively and literally, gave rise to concerns of equal sales representation in competitors' models.

## SELLING CIVANO

Although direct quotes were not released for *Inside the Civano Project,* Al Nichols Engineering, Inc. has spent time learning a great deal from Civano realtors about what it was like to be a realtor selling homes in Civano. The Solar Village had a unique combination of sales constraints and strengths. The community was being heralded as unique in Tucson with solar homes, open space, and alternative building materials. However, the absolute uniqueness of the project was not completely accurate. Veteran Tucsonans and real estate agents knew that there was a solar village that had been around since the 80s in the Sabino Canyon area. Silverado Hills, not far from Civano, was also unique as the first community on the east/southeast side that had considerable open space, areas called NUOS (non-usable open space).

By January 1999, little had been built in Civano. It wasn't getting off the ground as quickly as everyone had expected. In Tucson, new concepts had a history of taking a long time to grow from an idea to a viable and visible entity. Realtors with long time experience in the region knew Civano would be no different.

The buzz around the Tucson real estate community was that Civano was going to be different with regards to the community concept; design of the homes; and, even in the development and enforcement of the CC&Rs (covenants, codes & restrictions). There were many questions from potential buyers. Civano was going to be so different that potential buyers didn't know what to expect including what type of homeowner's association would be put in place. The Civano sales agents were told that it was going to be a friendlier homeowner's association. One that was different than people were currently used to.

The most successful realtors became familiar with the CC&Rs and the builder's design guidelines. The concept was indeed different and unique. Compared to a typical new home builder community, it was intriguing to some realtors to have the opportunity to sell something different, innovative, and new with regard to product and concept. In reality, the CC&Rs were not too different from other communities in Tucson.

Selling Civano required a working knowledge of the technology, the energy savings, the alternative building materials, and the community concept. The challenge was imparting that information to new homebuyers in a way that would create a feeling of confidence. It was important for the sales agents to help the buyers develop an understanding of the concept which would enable them to take that leap of faith to purchase in a community unlike any other in Tucson—with assurances that it would be beneficial to them then and into the future. What some realtors found disconcerting was the up-charge on the building materials and the cost of the photovoltaic solar. It seemed self-defeating. Today, the majority of the homes in Civano are stick-frame with only passive solar versus photovoltaic—much more like standar housing.

In addition, the concept of a municipal improvement district bond (MID) was a totally foreign concept to new buyers in Tucson. A municipal improvement district bond is a way of financing infrastructure improvements by passing some of the costs onto those purchasing the properties for which the infrastructure was improved. In other locales, MID charges are so common that the charges associated with the municipal improvement are bundled into the mortgage, in a single payment. In Civano, it appeared as a second charge which was difficult to explain to some customers. This was a challenge to the sales in Civano and kept many potential buyers from purchasing.

The sales staff was knowledgeable in the use of reclaimed water at Civano and used it effectively as a benefit to the buyer. Initially the cost of reclaimed water was lower than potable. The benefit to the buyer was obvious, that is, until the cost for usage was changed and the sales staff wasn't informed of the change. To make matters worse ... it was a homeowner who discovered the cost was higher and brought it to the attention of the sales staff. Sadly, the initial concept of gathering/meeting areas, orchards, and small businesses that were introduced to buyers was not completely followed through to fruition by the developer (at the time of this writing). It resulted in many angry and frustrated homeowners.

There was no doubt that Civano homes were higher priced. However, what successful realtors emphasized was that compared to a typical new home product, the homes in Civano were at least equitable if not better. Front and back porches were included. More insulation was included. These types of amenities were only included at an additional cost in other developments. However, it was a challenge for the consumer to get past the price point.

Civano sales agents reportedly had to understand and believe that the value was there. They had to know how to sell the benefits to the buyer of this type of community. They had to be knowledgeable of what a typical home's heating and cooling costs were, what the other home builders at other communities were selling—what their products were. Realtors had to know what each builder included and what a Civano home could offer with regards to energy savings for the present and the future. Civano sales agents were up against the price point on every sale. Buyers could go to a typical housing development and get a lot larger house with more yard if they did not care about utility bills. Buying green was not in vogue as it is now when Civano opened.

One of the major concerns regarding the profitability and success of Civano was that few real estate agents wanted to bring their buyers there. Real estate agents take on responsibility when they bring a buyer to a community and sell them on a concept that may not turn out to be what has been promised. It is very difficult for a community to be profitable if the real estate community is not behind it. Before Civano was even fully accepted by the Tucson community and the realtor community, a Civano builder went bankrupt. More than 100 of that builder's homes had already been sold—before they were built. This was a setback in gaining the trust of the realtor community in Tucson. Realtors risk being sued by their buyers if they recommend a community and it turns out to be something different than promised, or goes bankrupt.

The type of community Civano wanted to be when it grew up was an unproven concept in Tucson. The rumors about Civano were rampant throughout Tucson. It was said that it was elitist and/or that it was a community full of hippies where you would be required to grow your own food. Additionally, it was a far distance from shopping, food stores, and other activity centers. In contrast, a traditional production home community was built not far from Civano and became profitable in a shorter time period than Civano.

East and southeast Tucson was always known for its larger home sites. Buyers came to this side of town when they were ready to buy away from town and live on their acre of land to get away from it all. Thus, many potential buyers objected to the smaller home sites at Civano.

The initial sales concept for Civano was to be centralized under the developer's broker selling several different builders' products. This was another unique concept for new home building in Tucson. The broker's sales hiring strategy was to have only one experienced agent and the other agents would be new to new home sales. Some brokers felt that this was the best way to ensure that the new centralized home building concept would be successfully accepted by the sales staff. In hindsight, this decision was probably not a prudent move. An introduction of a brand new community concept such as Civano needed a full complement of seasoned and experienced sales agents to ensure the success and profitability of the community.

In the traditional newly built housing community, the site agent or sales staff is employed by the builder and they actually work exclusively for a given builder selling only their products. In Civano, agents were selling for all the builders. Resale agents are accustomed to selling several different types of homes by many different builders. Resale agents, as opposed to new home agents, are practiced in selling the entire inventory of homes available in Tucson—both new construction and resale. In Civano, instead of being totally dedicated to one builder and their product, the sales staff was dedicated to matching the buyer with the right home regardless of who built it.

When sales agents work in a new home community they go to the sales office on site and are there from when the doors open until the last person leaves. There are no regularly scheduled bathroom breaks or lunch breaks. When prospective customers come through the door, the realtor is expected to be up and it is time to sell. When it's busy, there is no rest. When Civano first opened there were times that at least one agent from the group had to remain on duty until 11:30 p.m.. For comparison sake, a typical new home community in Tucson would get 25 visitors in a week. Fifty or more potential buyers in a week would be considered a high count. Civano, on the other hand, was bringing in up to 250 prospects per week. That left very little time to write contracts or complete necessary paperwork. The sales agents at Civano had to work harder than agents in other new home communities because they were selling something that was very different than the average new home community and the visitor count per week was incredibly high due to the nature of the community.

The centralized concept proved to be very unpopular with the builders. The idea of the sales staff selling for all the builders made the builders feel that they had no control over the agents, whereas with the traditional model where realtors worked onsite for a specific builder, sales staff followed rules and procedures for only one company's products.

Reportedly, Civano wore some realtors out. Some left Civano very unhappy and it showed. Although it was such an exciting time to be able to be on the forefront of a concept that could take homebuilding and community development in a new direction in Tucson, within four months of Civano's grand opening the sales staff was told that they would no longer be centralized. The sales agents were told to find other positions, possibly with one of the Civano builders—if the builders weren't going to use their own sales agents. The alternative was to leave Civano. This took the sales staff by complete surprise. They were truly dismayed, discouraged, and frustrated after all the hard work and enthusiasm they had invested in learning the concept.

Builders spend company resources on educating their sales staff with regards to their products' unique features and designs and the sales agents, in turn, educate the prospective homebuyers. In a manner of speaking, the realtor is on the front-line of homeowner education—providing prospective homebuyers with the knowledge they need to make an informed home purchasing decision. Having a centralized sales staff may be a benefit to consumers and even the realtors, because the knowledge about all the different housing products is also centralized, but builders loose control over how their product is represented. In Civano, these dynamics led to the decentralization of the sales staff, and a very frustrating experience for many of the realtors involved with the project.

One more point of integration for a new community concept is to gain the support of the real estate sales community early and to consider a sales model and how it will affect the experience of the builders, buyers, and sales agents. Sales staff can only go so far to sell homes based on as yet unrealized visions for unfinished communities.

## GOODBYE FANNIE MAE

Fannie Mae and Civano were drawn together by the American Communities Fund—a congressional mandate for large development companies and funding entities to invest in innovative housing developments that revitalize communities, with a special focus on higher performance housing. Although the existence of this mandate did bring Fannie Mae to Civano as a project financier; mandate alone cannot transform an organization.

**McDonald:** Could you tell me more about the American Communities Fund?

**Kelly:** The American Communities Fund was a new fund set up by Fannie Mae. As they described it to me, it was a bit of a political fund. It was a fund that was trying to do good work in congressional districts, I think legitimately good work, to help garner congressional support for their company. But it was a new fund that was finding its sea legs and it was very sensitive to the internal politics of Fannie Mae and to the politics at large. So it was not what I would call a nimble or flexible fund.

**McDonald:** But what did Fannie Mae do for Civano?

**Kelly:** They were the minority investment partner and we were the development partner. They invested money with us to develop the project.

**McDonald:** But they eventually wound up with 100 percent of the project and sold off the remaining development to Pulte. When did you leave Civano?

**Kelly:** I left Civano January 2, 2000 at exactly 11:00 am (laughs). And I left because Fannie Mae and I did not agree with the vision of Civano. I believed that the vision of Civano was proven in the market place that phase I worked. Quite candidly, a monkey could have made a profit selling real estate in the late 90s and early 2000s.

**McDonald:** So why did Fannie Mae have a problem with it?

**Kelly:** Three fold. One, philosophically, we disagreed with Fannie Mae's vision of Civano. We believed phase I, with its different product types and technology, was an experimental lab that would help us determine how far we could push the sustainable envelope in the housing market. As the managing member of the LLC, and based on what we learned in the sales of phase I, we wanted to push the sustainable envelope further. Secondly, we believed that the project needed to be additionally capitalized to excel and we looked to our trillion-dollar partner for the additional funds. Since ACF was new, with little development experience, had no contractual obligation to invest additional capital, and was politically hypersensitive this request created some internal angst at Fannie Mae and eroded the relationship with the developer. The project decision-making ended up being penny wise and pound foolish. For a project of this size and importance—as we now understand today—it needed more capital to accelerate the business strategy. Third, the basic debate about smart growth in the country had really heated up around 1999 and 2000. In Arizona there were a number of initiatives on the ballot to stop growth. Therefore, Civano had gained a lot of attention nationally.

Due to the added national and political exposure of the project, Fannie Mae wanted to have greater control over the development as a way to protect its brand. As the developer and entity that had held more combined equity, mezzanine financing, and guaranteed bank debt in the project than Fannie Mae had investment in the development, we were not predisposed to give up control. That led to a conflict around who and how the project should be managed. Understandably, our relationship with Fannie Mae was new and they were reluctant to put their trillion-dollar brand in the hands of a small developer from Tucson, Arizona.

The Clinton/Gore administration was really trying to garnish support for the Kyoto treaty and they needed proof that the energy requirement that was in the world Kyoto treaty was, in fact, doable on a practical business level. So they turned to us as one of the shining examples of how the Kyoto Treaty would not unfairly burden the private sector. Then Governor Jane Hull, in 1997, turned to Civano to show how this was the best example of smart growth in the State of Arizona. So that third factor with Fannie Mae is that this project became highly visible on the political radar. So they wanted to be more involved rather than less involved.

**McDonald:** So did that lead them to buy everybody out?

**Kelly:** No, those three factors led us to realize that their approach to where the Civano project should go and our approach were not consistent. We didn't think that a trillion-dollar company was innovative, entrepreneurial, and nimble enough to react and really be the leader of what I would call an experimentation lab on innovative development.

**McDonald:** Why would they be left in full control of it?

**Kelly:** Because they are a trillion-dollar company! Fighting with them, friendly yet sternly, for a whole year to keep the vision in place—I finally sold out to them at a reasonable number.

Let me tell you the condensed story of Civano. We could literally spend hundreds of hours talking about this. There are plots, sub plots, and conspiracy theories. The lessons from Civano that are played out in these layers of conflict that you see everywhere laced throughout the story of Civano are really about the notion of sustainability. The old linear, polarized debate that has essentially limited any substantive discussions about anything in this country is played out in Civano. There is no 100 percent correct position. You look at the economic, social, and environmental impact and you weigh all of those issues simultaneously and you try to make the best decision you can. In that light, it requires everybody that comes into the equation to think differently.

I'll give you a true story. A guy named Henry Kelly who was the chief scientist for the Clinton/Gore White House was the guy who was basically responsible for defining the Kyoto Treaty standard. He is one of the guys who I'm sure provided some information for the Inconvenient Truth for Gore. He is a professor. He came out to Civano one day. He said they thought they were having Al Gore to come out and make some announcements about the environment, and they wanted to use Civano as the backdrop platform. About two weeks later, Gore's political handler decided they should do it in LA, because they could raise $2 million, so they didn't come. But that is subsequent to my story. But before he was going to bring the vice president out, Henry asked, "Kevin, what's really going on out here?"

I said, "Henry, the solar people don't think there's enough solar, the permaculture people don't think we're saving enough native vegetation, the conservationists think the water savings should be more stringent, the alternative builders think we're too traditional; the traditional builders think we're too alternative; there are 27 community groups and government agencies that feel they have a voice in the definition of Civano—meanwhile, my bankers don't understand why I'm telling them I have to go meet with a community group before I can go sign a legal document for a personally guaranteed loan and my other investors think I'm way the heck out of the box." I said, "Basically, Henry, everybody is a little bit angry at me." And he smiled that smile of a person who had seen his share of battles, and he said, "Kevin, then you're right where you need to be."

That, to me, is the summation. The people who came to Civano and said it was only about solar or only about neo-traditional design, or only about mixed use, came to Civano with the old polarized linear world that it's only one cause that has the preeminent decision, then they are lost. Like the developer, like me, who says, "It's only about me because I hold title to the land. I have the power and I pay the mortgage." Every person who believed that their own position was the only right or righteous position, including me, soon learned that in this magical, electric place called Civano, you needed to think differently in order to really learn and get the lessons from Civano. The lessons of Civano were profound and extend well beyond the mechanics of master-plan development. All of the stakeholders who were involved with the project, albeit the developer, investor, solar community, builders, land planners, government, etc., thought they were there to help the other constituencies learn from their knowledge. But sustainable thinking teaches us about interconnectivity and interdependency.

So you come away realizing that your stake is no more important than any other and that we need all of us to build a better world. The world cannot solve its complicated problems by limiting the discussion to who is on the right or left, or pro-business verses environment. This is an interconnected environment, society, and financial and geo-political system so we'll all be better off when we approach our problems through a prism of interdependency.

John Laswick carried the responsibility of bridging the traditional way local government saw its role, which was as the "policemen" over the private sector verses truly being in partnership. It was a tough role to be in but John managed it better than most; surely, better than I could have done.

John Laswick was the City of Tucson's project manager for Civano and also witnessed the dynamics of bringing a huge financier to Civano, one with old ways of conducting the land development business.

**Laswick:** Well, Fanny Mae was never comfortable selling it as a sustainable new urbanist community. In fact, I'll tell you something that will shock people that Kevin Kelley told me. When they originally got their commitment from Fanny Mae, I can remember going out to their office and celebrating with them. Well, I was out there for something else and everybody was really psyched about it, that they had gotten this financial commitment. And he said, "You know what they told us, they said don't mention sustainability and don't mention new urbanism." This was before we even started.

And the problem was that the guy who had gotten it approved was not at Fannie Mae anymore. So we were back to just sort of conventional real estate people. And the American Communities Fund, which was the source of the equity, had been set up right then basically to finance inner city redevelopment. Everybody understands that on the east coast where they've been doing that for 60 years.

In retrospect, it is little wonder that when Fannie Mae took over the Civano project they did not promote the community based on sustainability either. The terms green and sustainable are non-explicit terms at direct odds with liability concerns, as well as ideas and perceptions about affordability. Jim Singleton, a city employee who helped with Civano, bridges the gap between the questions of affordability and marketing green building in his interview with Al Nichols' intern, Brian McDonald.

**Singleton:** I'm Jim Singleton. My early involvement with Civano was while I was the building official for the City of Tucson. I retired in January—January the 8th of 1993. And during the late '80s or early '90s this proposal for this Civano project came to me. Al Nichols, a local mechanical engineer and energy conservation engineer was very actively involved in getting the project started. What was developed was a proposal that went to mayor and council. I don't remember if it went to the board of supervisors or not. But I know it went to mayor and council to authorize rezoning for a specific subdivision that would be very energy conservation conscious. And the subdivision structures that go in out there would be required to use both solar and alternative energy conservation measures to conserve both energy and water.

Obviously there is a little bit more cost involved initially up front in adhering to the Sustainable Energy Code versus the IECC. However, the builder has a sales advantage in that he can promote that he's going to save the owners on their utility bills throughout the life of the building. And the owner then is the one who really benefits from it, even though he probably pays a little higher cost up front for the building, the structure, to start with. But in the long run, he saves on his energy bills. And if I remember correctly, Al Nichols indicated that his estimate was that they would probably recoup the difference in about three years just on the savings of the energy.

One of the things I remember that they advertised starting up Civano was that this was going to be a community where they would have the jobs available in the community and they were going to try to build it so people didn't have to drive somewhere to a job, they were going to have commercial activities and their jobs that would be available there in the community. And supposedly some of the people there could work in the community and walk and not have to drive cars all over the place. And it would conserve additional energy in that respect. Well of course back then the price of housing wasn't quarter-million-dollar homes, when first started. The price everywhere has escalated and gone up.

And this issue of affordable housing, that's a—that's a term that is bandied about quite often. The last president of Southern Arizona Home Builders bandied this affordable housing thing around for a long time, so did the politicians. And one day the vice president was standing up there; and he said well, when we say affordable housing, you have to understand it's affordable to the group that we're marketing to. Okay. So—that puts a whole different thing on this affordable housing. See, if you're talking about building minimum housing for starters for underprivileged and people that don't have hardly any money, then you're going to build a totally different kind of house. But if you're building affordable housing for middle-class America that can afford quarter million and half a million dollar houses, then you build that style of a house, it's affordable to that group that you're marketing to. If you're building affordable housing for millionaires, then, you know, you build appropriate to that group that you're building the affordable housing for. So affordable housing, you've got to be careful how you use that term.

# Defining Affordable Housing

According to the U.S. Department of Housing and Urban Development (HUD): "The generally accepted definition of affordability is for a household to pay no more than 30 percent of its annual income on housing. Families who pay more than 30 percent of their income for housing are considered cost burdened and may have difficulty affording other necessities such as food, clothing, transportation, and medical care." So, in practice, affordability is a moving target. Odd, though, that this assumes that a family making more income than another also will need more money for necessities such as food, clothing, transportation, and medical care.

The truth is that we each have the same basic necessities, with some variations. There are also more extreme exceptions for which the welfare system was developed. So, if it takes 30 percent of the income for a family of four earning $100,000 for housing, and the rest to provide the other necessities of life; what does this say about a family of four living on $50,000? They simply need less?

Using today's mainstream rhetoric, "affordable housing" is simply another way of saying "the most any given individual or family can afford for housing." For the family earning $50,000 annually, affordable housing may not include double pane windows, a high efficiency heat pump, water efficient yards, and a super insulated building. In Tucson, it may mean they will purchase the ubiquitous 1960s double brick cottage with no roof insulation and drafty metal-framed single pane windows—or the newly mass-produced tract house that barely meets code requirements. The irony is that affordable housing for lower income families requires that they spend a much greater percentage of their remaining income on things like utility bills and room air purifiers, repairs, and perhaps even medical bills. So what happened to "… affording necessities such as food, clothing, transportation, and medical care?"

Furthermore, poorly constructed housing ends up becoming a waste management problem sooner than well-constructed buildings, which equates to a cost for society as a whole—the immediate effects of which are arguably weighted toward those with fewer means. Of the two cheaper homes mentioned above, the older brick home and the production tract home, many people would have no qualms with Tucson's double brick ranch homes. These, for the most part, are very robust structures and last many, many years. Retrofit and upgrade of a building with an excellent structure and foundation is a much more attractive option than trying to maintain a lemon. Retrofit can be an expensive proposition that is usually regulated to those who, once again, can afford more. This type of affordable housing actually is only cheap housing at first, and then quite unaffordable thereafter.

The family with a little more income, on the other hand, may find their version of affordable housing costs a little more up front to ensure that the house is truly affordable for the long run. Their home will stay out of the landfill longer, it will have a better chance of being remodeled when the time comes, and it will be less of a burden on the family living there. It will be an asset to the owners and less of a liability to the rest of the world. So, the term affordable housing should include the ongoing cost of ownership as part of the equation. Also, instead of defining affordability of a necessity like housing as something that is scalable and tunable to what the customer can afford, we should consider defining what the basic house should do for its occupant, at a minimum, to be healthy, efficient, and viable for a long time. Then the cost for providing such a structure could be determined given the location where it would be built, and that number would determine what constitutes affordable housing.

Then the questions would be about who can afford such a home, what the deficit of those who cannot is, and how the deficit can be bridged. One answer may be subsidies, just as happens now in HUD developments. No matter how we got people into those homes, the end result should be that poorer working classes are not being further challenged by only having access into affordable housing that zaps their remaining funds or leaves them in uncomfortable or unhealthy environments. Even if a family needs some assistance purchasing a home, if that home is efficient and costs little to own and operate, the family living there is much more likely to remain self-sufficient on an ongoing basis. Families living within their means in a healthy environment, in homes which pollute less and create less waste, is good for America. To achieve this, base-level affordable housing should also provide a base level of efficiency and affordability that is better than what is minimally required today. Although the International Energy Conservation Code (IECC) has been advancing significantly in the last couple cycles, it is still up to the local governments to have the wherewithal to adopt these (or more advanced) codes and enforce them.

## COMPOUNDING EFFECTS OF ENERGY EFFICIENT BUILDING ENVELOPES

Even if all the homes in our nation are retrofitted or built new to a very high level of efficiency, not every person will afford huge solar panel arrays or ground-source heat pumps. But it is very much within the building industry's capability to ensure that all new homes have an efficient envelope to begin with, incorporate basic cost effective efficiency standards, and provide gray water and solar energy connection points. However, if we make such requirements on homes (and commercial buildings) the value and positive impact of more expensive things, like solar energy, go up for everybody. Why? If a home uses less energy to begin with, one solar panel provides a greater percentage of that home's energy requirement. So it will require a smaller investment in alternative energy to provide the needs of that home.

The question of affordability strikes deeply into the green building movement. While the terms green and sustainable are at direct odds with ideas and perceptions about affordability, perceptions about what makes construction green and sustainable affect how people determine value. As mentioned earlier in this text, there were models of architectural successes, master-planned communities that became highly sought-after places, even without any mention of green or sustainable in their marketing. What happens when somebody actually comes along and creates a greener and more affordable home?

What was found out in Civano is that efficient and affordable housing in a good community will appreciate quickly. The affordable units (based on Tucson's mean household income) in the highly-touted first phase of Civano quickly appreciated beyond "affordable" status for the majority of middle-class Tucsonans. Martin Yoklic discussed some of these aspects with Brian McDonald, and some of his words follow.

**Yoklic:** Livability and resource efficiency are desirable aspects for communities. Village Homes in Davis, California, a reference model for Civano, provided an illustration of how community affordability is impacted by these factors. Recent data on housing values in Civano confirms that well-performing housing (energy and water efficiency) in livable communities increases in value faster than communities without these attributes.

There are two contributors to this story, the developer and the affordable home buyer. For the developer, maintaining profitability means building in the way that has proven cost effective and marketable. For the mean income buyer this means finding housing that fits their budget. While this matching has been effective in promoting economic growth, the implications on community livability, resource efficiency, homeowner equity, and housing industry viability have begun to come into question.

One of the objectives of the Tuscon Solar Village Project (Civano) was to demonstrate that energy and water efficiency and good community design could be profitable for the developer and welcomed by the buyer. In concept, while the developer's cost efficiency is impacted by additional design/engineering, approvals, and technology within the housing, these are offset by higher density, reduced street width, more efficient land use, and product marketability. For the buyer, the smaller housing and reduced utility costs help make the product more affordable.

**McDonald:** It seems like builders have been doing it almost the same way for a long time. How much did Civano challenge them?

**Yoklic:** From the developer's perspective the project has been difficult. But Civano has shown that efficient community design and resource efficiency in housing are desirable in the marketplace. We are increasingly beginning to realize how important these goals are, and will continue to be, to the economic vitality of the housing industry in the future.

From the buyer's perspective, the growth equity of their homes has outpaced other developments in the region, pushing the housing value in Civano beyond affordable The bad news is that Civano is not as affordable as was intended. The good news is that those who purchased when the housing was affordable have gained more than if they had bought a new house elsewhere.

It is time to take the lessons learned from projects like Civano, Village Homes, and others, and bring them into a new model of resource efficient community development.

Given the information gathered from watching Civano, and hearing the opinions of individuals involved with the project, it stands to reason that if every house was built better; then better houses would be more affordable for everybody. If better houses remain a unique product, they will always be expensive and out of reach for many people because of simple supply and demand economics. On February 22, 2009, there were 10 homes represented in Arizona on the Listed Green Web site (http://www.listedgreen.com - Listed Green is a USGBC member company, business member of Co-op America, and member of Global Green USA.) The average asking price for these homes was $563,900, the median was $412,000, the maximum was $1,300,000 and the minimum was $210,000.

Given a 30 year fixed-rate mortgage, 10 percent down payment, 7 percent interest on a loan, and incidental ongoing property tax and private mortgage insurance costs; combined with the definition of affordable housing as using 30 percent of a family's income—the following is true about this small sample:

- The average priced green home ($412,000) would require a family income of approximately $120,000 per year to be considered affordable.
- The most expensive home ($1.3 million) would require an income of approximately $380,000 per year to be affordable
- The minimum priced home would require approximately $62,000 annually to manage affordably.

The next most "affordable" home above the $210,000 asking price was $329,000. These certainly do not represent the only homes in Arizona with various environmentally conscientious design elements. If Pulte listed its homes on this website, the average price would have dropped significantly and the lowest price home would have been closer to $120,000. However, this sample does represent diverse approaches to being greener such as on grid, off grid, planned community, remodel, and luxury homes from across the state. What is perhaps most interesting is that the least expensive home may be the greenest of all the options. It is the only home in the group that is located in a pedestrian-friendly location in close proximity to the town center. The $210,000 home is also one of only two homes remodeled with the environment in mind rather than built new, which requires new land, new materials, and all the other resources associated with new construction.

Given all the diversity of housing on the market, the disparate views of what is affordable and what constitutes green design, it is understandable that builders and developers intent on providing clear and honest advertising may decide to avoid heavy green advertisement approaches. Once again, we come to the issue of education. If homeowners and builders alike understand the features that contribute more to a sustainable lifestyle and housing technologies, then builders can simply list the attributes of their homes factually, without greenwashing. The consumers will apply their own perception of what is green toward making their home purchasing decisions.

The first rule to increase sustainability and efficiency is to conserve. Use less, use smarter, walk lighter and waste not, want not. The generation that lived through the Great Depression possesses a memory of economic pain that is more vivid than most who did not experience that widespread American catastrophe. What good came of it? New ideas, a conservation ethic, hard work, camaraderie, and the New Deal could all be considered fruit of the depression. When times get difficult, people tend to adapt in ways less considered during times of plenty. But when there is plenty, as if we are looking at bright shiny gold, our eyes are dazzled by the glitter and many of us forget what it is like to be without. Some still living with memories of the Great Depression, our parents and grandparents, remember it is better to use less and save some for later. Some of their kids remember the ethic. During times of less, these values often shine through.

Ideally, those individuals and organizations in touch with the wider consequences and concerns about developing the built environment will continue to create a doctrine of conservation that is derived to cater to the needs of people living in a poor economy and through rough times, both now and in the future. Planning for the low points in our economy would predicate the construction of places that are inexpensive to operate, maintain, and keep so that people working with less will be able to maintain a higher quality of life. These same measures will help ensure that during good times, the same people will have a greater chance to build wealth.

# 7

# GROUND BREAKING

## *Neighborhood One*

> *Every step of progress the world has made has been from scaffold to scaffold, and from stake to stake.*
>
> —WENDELL PHILLIPS speech, October 15, 1851

Metaphorically, the process for envisioning a large-scale master planned development is like looking at earth from space, and then zooming in. Initially, one can see the master plan: a massive sphere that contains large scale features, shapes, and their spatial relationships. From this perspective, one gets the big picture but with an averaged, generalized view of finer grain details like color, material composition, structure, and general functioning. Upon moving in closer, the subsequent levels of detail come into focus: continents, jet streams, countries, mountains, states, counties, rural, lakes, urban, rivers, buildings, concrete, purple, brown, yellow, streets, paths, scaffolds, and stakes. Move in even closer and we see the individual building design packages, framing plans, building permits, soil type, microclimate, sewer lines, energy calculations, and low flow lavatory fixtures—the complexity.

From the moment of the first world-view conceptualization of Civano to the time the details were planned enough to allow ground-breaking, 15 years passed. Many of the processes and ground-laying that had to be completed for Civano to become a reality were being developed alongside the project. Many of the ideologies which early innovating projects like Civano had to nurture around them now exist on a wider scale and independent of specific projects. The Leadership in Energy and Environmental Design (LEED) building/project certification process is an example of what can happen in an environment primed for high performance construction practices. Although much of the politics involved when collaborating with the jurisdiction having authority on a large project have their own processes (rezoning, for example), programs like LEED streamline the integration of environmentally conscientious design and process details into the project.

The first Civano neighborhood has more detail at the urban design level than the rest of the development. There are more integrated design features in the urban landscape and the buildings themselves. Homes and community buildings were built from many different materials using several different processes and all placed in a setting stylized

**Figure 7.1** Row of homes in Civano with solar and high performance envelope. *Drawing courtesy of Justin Cupp.*

to fit the Sonoran Desert in form as well as function. The end result is a tapestry of environmentally conscientious building designs constructed within a fully engineered site. The variety of designs was intended to make Civano a showcase of more sustainable building approaches. This chapter describes some of Civano's urban setting, explores aspects of the IMPACT process more deeply, and introduces some of the different types of materials, building types, wall systems, solar devices, and construction strategies used in Civano's flagship, phase I. Figure 7.1 is an artist recreation of the view facing norh from the roof of the Al Nichols Engineering, Inc. live/work facility.

# Urban Design Features Specific to Phase I Civano

This place is colorful![1] There are landscape canopies along walking routes of Mesquite, Palo Verde, and Acacia trees. Walking beneath is comfortable both physically and emotionally because of the shade and the knowledge that these native plants use little of the city reclaimed water which feeds them. The mountains beyond are visible from most street angles. It is not like being trapped in the concrete valleys of a city or blanked out by tropical rain forests. The community is uniquely desert bosque.

---
[1] To view some images of Civano, visit www.civanoneighbors.com.

## THE STREETSCAPE AND LANDSCAPE

Civano's neighborhood I has roads that scale from larger two-way streets with fully planted medians in the middle, down to walking paths in the landscaped desert country areas—and everything in between (see Figure 3.1). The roads curve in opposite directions outward from the neighborhood center almost like the spiral arms of the Milky Way Galaxy. Sloped curbs in most of the project are not only bicycle friendly, but they allow for large vehicles to cross paths more easily on the narrow, black-topped streets because they can each get a wheel up on the curb without hazard. Where roads do not go, paths, sidewalks, and trails connect pedestrians to locations across the community.

Looking at a map of neighborhood I, one may recognize aspects of the first walking cities which mostly developed outward from a civic and/or merchant center, like spokes from the hub of a wagon wheel. However, whereas in many cities fringe development happens largely unplanned and the city adapts and grows into it, neighborhood I is obviously designed and deliberate. The paths are built, not worn in by feet; and the landscape is planted by hand, not chance. Still, compared to the vast majority of urban housing projects, neighborhood I's streetscape and trail integration make it feel quite organic and close to nature.

There are road planning issues that could have resulted in better access. The exits from the neighborhood I community merge into the main city thoroughfare of Houghton Road. There are no traffic lights planned for either of the two immediate exits of neighborhood I. This means that during rush hour long lines of cars traveling up to 55 miles per hour can make the Civano residents' wait at the stop sign lengthy. If only the northern most road of Civano had been extended along an existing haul road which terminates at a traffic stop light, perhaps 100 yards further than the current most northerly exit, there would have a been much safer and more reliable access to and from that part of the community.

There were mistakes made by the combination of high density polyethylene (HDPE) water piping and improper backfill techniques which resulted in a series of leaks, what some neighborhood I residents are calling "Civano springs." These "springs" are the failing water line where it branches out and connects to meter boxes for individual homes. The plumbing was set in trenches and backfill was compacted on top without care taken to fill beneath the HDPE piping first. HDPE is very resilient and rather than breaking, the material stretched and deformed with the compaction of the road bed. Eventually the HDPE on some of these connections has given way at the fitting where it was most stressed. When breaks occur, water flows along the shortest path it can to the surface. Presto, a "Civano spring!" Since the City of Tucson is responsible for maintaining the streets, they have made a science of cutting the asphalt and patching it as "Civano springs" occur.

Part and parcel to the streetscape is the landscaping. The landscaping is an integrating factor across the development and is a large portion of what builds a sense of place there. The trees that were planted in the rainwater-collecting median areas serve multiple purposes. They provide shade and a buffer for pedestrians and they shade roads to reduce the heat island effect. This has a positive cooling effect for the Civano microclimate. Also, perhaps an unplanned effect, the landscape plants grow out and squeeze the automobiles a bit by making roads seem narrower, which encourages some motorists to slow down as they drive through. These large plants, watered with city reclaimed water, are also habitat for many species of birds found in the desert. The birdsong in the neighborhood is delightful. Amazingly, these transplanted trees are so strong and healthy that in some areas they have heaved sidewalks.

The amazing good fortune and hard work of saving many hundreds of mature trees for replanting in medians came with some unforeseen side effects. It may have been 50 years before small saplings that would have normally been planted in such a site could be able to lift several segments of sidewalk and crack masonry block walls. Much of what was done in Civano phase I's landscape made the place feel like a community that had been there for a while, but it also introduced some of the maintenance concerns of an older community.

There are two main green spaces in neighborhood I for children to play soccer or for outdoor gatherings, such as wedding ceremonies or picnics. These grassy areas are watered with city's reclaimed water and serve as some of the anchor points for navigating through the neighborhood. The trailed areas in between, called the desert country areas, provide an inter-development safe haven for pedestrians or hikers who like the desert flora and fauna. There are decomposed granite (DG) paths which are maintained by the HOA throughout these landscaped areas—usually well enough groomed that a wheel chair could make its way down the trail.

Some of the more unique new urbanist techniques were deployed in this area to face homes toward the desert country area—front doors toward the trails. The idea was that pedestrian-minded residents would leave by their front doors and enter the walking trails to commute to other areas in the community. Although the trails are often used for walking, the vision was not fully met and most residents view their alleyway as the front of their houses and the patio facing the desert country trails as the back.

## The IMPACT Process, continued

Whereas Al Nichols Engineering, Inc. has been responsible for monitoring the energy and water consumption of the Civano project, Gal Witmer has been the individual primarily responsible for monitoring most other aspects of the Civano project pertaining to the development agreements. Excerpts of an interview with Gal follows, intertwined with portions of the most recent monitoring report she wrote.

**Witmer:** I'm Gal Witmer and I'm an architect here in Tucson. I got involved with Civano when I worked for Swaim Associates. Sometime around 2001, we had been granted a commission through the Vail School District to design the Civano Community School. So we worked with the school for a couple years to get that rolling. That's where we met Al Nichols, Judith Kilroy, Lee Rayburn, because we had to go through the review process for the school. At the culmination of that project, Fannie Mae approached us to take over the reviews, the third party certification to ensure that the houses and all the buildings being built were meeting the criteria of the IMPACT standards and the Civano master plan, things like that. So that was how I got involved and met a lot of key people from Civano and I was basically the person at Swaim that knew what was going on with Civano. It was a very interesting project.

From there, after I left Swaim Associates, Phil Swaim had me continue to do Civano work and I stayed involved and continued to do the monitoring reports, to write the reviews. When Pulte purchased the project, they approached Swaim to continue the monitoring and Phil told them I was out on my own, knew what was going on, and so I'm now a consultant for Pulte, still doing the reviews and writing the monitoring report. It's probably been about eight years now.

**Laros:** What is involved with the monitoring report? How do you do the report and why do you do it?

**Witmer:** Basically, it was written into the master planning documents, the PAD, and development agreement documents. They realized that they needed to track if they were meeting their goals, how they were doing. If these are the goals that we're setting for ourselves, are we meeting them? Are we progressing? Some of the IMPACT standards had certain goals that were initially 10 percent that they wanted to gain on a certain area, and then they increase. So when they get to a 50 percent build-out, the same goal may move to a 20 or 30 percent target.

When I took over writing the monitoring report from Lee Rayburn, I used his format, or some of it. As I learned more about it, I took some stuff from Lee and put our own spin on it. It's all there, in the PAD, how to do it. And that is the format I've been using ever since I was at Swaim and I still use today. Over the years I've tried to find ways to do things a little different or research a little heavier. I've recently incorporated a survey, sort of a grass roots survey, but I got that out through the HOAs and whatever means I can to try and find things out. Like, how many people are buying hybrid cars? How many people are working from home? How many people are using recycled products?

We know how many people started off with xeriscaped yards, but we don't always know down the road if that changed. You guys study it so you know if we aren't meeting the water standard but we won't always know why. Is it because people are bringing in other plants? So I'm trying to get at the root of that. It's limited, but last year we did really well.

From wall systems to trash abatement goals, the small details of Civano are numerous and, at times, complicated. The landscape must compliment the use and intent of the project, but has its own needs as well. The streets need to convey cars, but they must also drain water away from structures without washing away landscaping or damaging the hydrology of downstream ecosystems. The people need to get rid of their waste, but the community goal is to increasingly recycle. All of the small details about how a place will function are tied intimately to form, but how does this relationship play out? Martin Yoklic discussed in an interview how integration of some of the site features can compound the benefits of functional form. Figure 7.2 is an artists' recreation of a water calming feature in Civano.

**Figure 7.2** Water calming features such as gabions are strategically placed throughout the community to handle runoff from pedestrian paths and overflow from medians and lots. *Drawing courtesy of Justin Cupp.*

**Yoklic:** One example of integration is making our drainages so that you could see them and walk through them. Not trapezoid sections behind the project that engineers designed to move the water off the project. I mean, that's just lack of integration. The development community takes the plan, says, "Here, engineer, I want to get this many lots on there, deal with it." Engineer says okay, I'll do something like this and this drainage has to go here and looks like this. And if it's a trapezoid section, it can move the water faster and quicker and easier off the project. Developer says okay, I got my lots, I'm gone.

And here we asked how we make that water an integral part of the project. And it got absorbed, to the dismay of the county reviewers—or the city reviewers.

**McDonald:** Why the dismay?

**Yoklic:** Because they wanted to see the trapezoid sections, that's what they're used to.

**McDonald:** You know, the drainage thing, that is an interesting point. Was Civano based off of the Davis model?

**Yoklic:** Well, yes, except Davis had another problem. It deals with drainage from a totally different perspective. There's flooding problems in Davis. It's a low place. Low and flat. So they made the common areas flood and not disturb the residential lots.

**McDonald:** They built retention basins?

**Yoklic:** Essentially. Now they won't say it that way, but that's essentially what it is. Because it can—it can take a ton of water and use the water as opposed to just shooting it off. But in a sense, we do that here too. I mean, we still get it off the property, but it does something on the way hopefully. There were compromises. There was a lot of back and forth doing that charette thing when we were doing it—because one area was modeled after the Davis model. The area where you have the feeders to the houses, an alley, and then it goes into the garages, and then it dead ends in a cul-de-sac, and then the front yards are common space.

Anyway, that's the Davis model. We have to do things differently here because we have a different set of constraints—essentially what happened when the folks got the project, they couldn't make it pencil out with the density it was in. So they increased the density through the incorporation of new urbanism. That was a vehicle to get the density up. I don't know whether it was strategic or socially responsible or whatever you want to call it, but it allowed the density to increase. And it created a really unique model.

**McDonald:** The grading allowed density to increase?

**Yoklic:** The grading? They didn't do a ton of grading here.

**McDonald:** I've seen some interesting pictures of the way they removed all the trees, dug three feet deep, and pushed all the dirt to one end. Dug three feet deep—and I would assume that that's very costly and very extensive.

**Yoklic:** No. That's the way they do them all. I don't know if you have done any engineering work or anything like that. But when you do projects like this, there's a minimum acreage that you don't want to—you don't want to disturb the site, especially if it has any—any terrain at all. And it—that usually occurs with anything under about a half an acre you have to grade to get the drainage to work because you're covering a third of the space with your building.. The building has to be higher up so the water has to go away from the building. And it can't infringe on the neighbor's property, so you may have to make provisions to it to go somewhere between. And to get that all sorted out, you have to grade.

**McDonald:** But if we all have this integrated approach, aren't concerns about infringing on the neighbor's property counterproductive? If you're using it together.

**Yoklic:** Well then—then it goes to liability. By the municipal laws, you can't put your storm drainage on another person's property because then you're affecting their capacity to do the best they can with their property. So there's this other social order that we have to work in. And we really pushed them on this project to get it done the way we got it done. I mean, just doing the drainage that we did here was like major mind-blowing experience for the people who reviewed the plans.

There's the place right up the road here when you go to the top of the hill. There's a nice little development there, houses going back to terrace, pretty dense, that roll with the hill. And then—then it's cut off, and there's a street going down with houses on both sides of the street. They moved thousands of yards of dirt so they could get the houses matching side-by-side. And the engineer just said yeah, we can do it. And they got this great big cut wall. Now all they had to do was make fewer houses and just do it on the hill. But they wanted to get all those houses in. So when you put density in, you have to grade. That's—that's just the way it goes. You have to get the water moving in the right direction.

I just finished a study last January, we talked about it a few places. This is related to the water issue in the region. Do you know the little fact here, if you take the urbanized area of Tucson, and you calculate the amount of rainfall that falls on it, and you compare that to the municipal demand of that same area, it's nearly equal?

**McDonald:** So if everybody's water harvesting, problem solved.

**Yoklic:** I'm not saying that. There are barriers to that. You took this huge leap, you know. What the study shows is that we could, with some imagination, some regulatory overlays, some incentives, creative development, some stuff like that, we could reduce our municipal demand by at least half just by using the water that falls from the sky.

We don't use any of it because there are arcane laws that prevent us from doing it. We aren't allowed to drain on other people's property. We also aren't supposed to hold water upstream for downstream users. See, if we use it up here, they don't get to use it. So that's against the law.

So there're all these overlays of dilemmas. And there're the navigable rivers. If we hold it up here, you can't run a boat on the Salt River. Well, they haven't done that in a long time anyway. But those—those ancient laws or arcane laws have caused us some problems. And we're going to hit the wall—we're going to see that start.

I developed a system for a product manufacturer in Madison, a demonstration on a house that had the house set up to take the rainfall. The rainfall goes into a small tank, and there is a little pump on it and a sensor on the tank. And when the tank has got water in it, it's used to flush the toilets and do the washing. And when the tank's empty, it has a sensor on it, has a little relay, switches, automatically switches to the potable water source. I designed it for the Arizona climate because we often don't have water in our tank. But when there's water in our tank, it automatically uses it for flushing the toilets and such. You need a little bit different plumbing system. You need a tank and a system in the garage. Now when I was doing the research for this, you can buy the pump, the relay, the controls, and stuff like that in a nice little package that you mount on the wall and hook all the pipes and sensors to in Denmark, off the shelf. And we can't buy it here. And I can't get one for here because it's a different voltage and they won't distribute it here because they don't want to deal with liability issues.

I mean, we could do it here. It's not a big gizmo, it's a pump and a relay and some controls and some solenoids and some backflow preventers, and you put it all in a box and you got it. It's a done deal. And you can put it right here, the whole thing, the tank and everything. Have your wire coming in, have your pump. You know, in your garage where you have your hot water tank and your heater, just have one more appliance there. It's an appliance for the house. It's like the solar hot water, same stuff. I mean, it's all available to us. We just need to figure out how to do it.

There are many, many approaches to building more efficient and comfortable housing for people. There are numerous wall systems, insulation types, plumbing strategies, and mechanical equipment designs. Determining which ones are best is not necessarily as straight forward as choosing the highest efficiency product. A holistic approach requires that all the attributes of the materials and methods used are considered in unison. Especially in the case of Civano, intended to be an example of how production housing could be made much more earth friendly, the methods, materials, and equipment used were ideally items that would be replicable outside of Civano. Some were, some probably were not.

## Building Materials, Techniques, and Technology Adoption

Gal Witmer talking about the Civano community survey she does as part of the IMPACT monitoring report.

**Witmer:** We had about 140 people return surveys, which is really high. Actually, based on the results I saw this past year, I revamped it for this year. I realized that a lot of people really didn't know what things were. I'd ask, "Did you add a photovoltaic system?" And some people would respond that they had one that came with the house. I would realize that they were thinking of solar hot water. I had people who would call me on the phone, not even knowing what it was. So there is a need for education.

This year, I've added images with clarifications. A picture of a photovoltaic system is provided with a statement that says, "This is where you generate electrical power." So I've tried to make it more educational, more user-friendly. I want to actually take a step to get it online. That will be a project I'll work on this year. Maybe for next year I'll launch it digitally. That way, we could get people to fill it out on line. What I have noticed is that there are a lot of citizens in Civano who will continue to use the mail or the fax and so there will always be that paper usage.

## Solar Power in Civano Phase I

Civano's first Sustainable Energy Standard

Excerpt from: **Impact System Monitoring Report #13**

Community of Civano Neighborhood 1 & Sierra Morado For the Year 2007
Gallagher Witmer, Architect LLC, Distributed: July 21, 2008.

### 3.4.6.2 Development of Solar Resources

All new construction must demonstrate prior to building, their intent to provide "beneficial solar" in their projects. Possible choices being:

- Thermal or solar electric space heating systems
- Trombe wall or clear view collectors for space heating
- Solar photovoltaic systems
- Solar thermal/electric power generating systems, including stand-alone and grid connected parabolic trough and dish stirling
- Solar day lighting systems specifically designed to capture and redirect visible solar energy while controlling infrared energy for at least one half of the non-bedroom space
- Passive building heating for the winter through the use of optimum window shade structures and orientation
- Solar water systems for domestic water heating or space heating
- Solar pool or spa water heating
- Solar oven that is built into the structure
- Solar food dehydrator that is built into the structure
- Solar water distiller attached to the building.

Many of the residential projects utilize at minimum solar hot water heaters to achieve this requirement, with some homeowners installing additional photovoltaic panels for electrical energy. All homes in the community are required to be outfitted for future solar hot water use if a solar hot water heater is not provided. Another requirement of the review process is that structural calculations are provided and an area of the rooftop is allocated to allow future installation of photovoltaic panels on all properties.

Changes were made in the past years updating the Sustainable Energy Standard. Per the new standard, "The calculated target annual energy consumption of the building lighting, mechanical system, and domestic hot water heating shall be less than 50 percent of the energy required by the ANSI/ASHRAE/IESNA Standard 90.1 2001 without amendments for the purpose of calculating the minimum base case, otherwise buildings must also meet the adopted International Energy Code of this jurisdiction. In addition, the minimum displacement goal of energy by solar devices is prescribed as a function of residential bedrooms at 550kWh/br/yr." This new measurable standard can be prescriptively met using a typical solar hot water heater for up to four bedrooms.

With the recent rise in energy costs, solar energy is becoming more attractive. Many tax incentives and rebates exist to assist homeowners. As energy prices climb, the paybacks and justifications for utilizing solar energy increase.

## SOLAR ELECTRIC

A photovoltaic (PV) solar panel as seen on the left of Figure 7.3 coverts solar radiation directly into electrical current. These panels produce direct current (DC) which can be

**Figure 7.3** Solar photovoltaic (left), hot water solar collector (top right).

used to charge batteries or to power DC equipment and appliances. Alternating current (AC), which most American homes use, is created by sending the DC electricity delivered from either PV or battery banks through an inverter. The basic mechanics behind the functioning of a PV cell occur on the atomic level and depend on some of the characteristics of semiconducting materials. Most PV cells used today are made from silicon but other materials are also used, especially for thin film applications.

PVs work based on the ability of solar radiation to excite electrons in the semiconducting silicon. When photons from the sun hit electrons in the silicon, the electrons try to elevate to a higher valance. Valance can be thought of similarly to an orbit. When an electron gets energy input (excited) it can move to a higher orbit around the atom's nucleus. However, in a photovoltaic panel, there is a conductor that captures the electron when it tries to elevate to the higher valance and carries the electron away. This leaves a hole where the electron once was.

Meanwhile, other electrons already set in motion around the PV cell's conductors are drawn toward the holes that were created as others left. All the electrons go in the same direction due to the use of diodes in the solar cell design. The sum of these interactions across many cells in a solar panel adds up to create direct electrical current. This kind of solar energy production is possible only in materials with electrons that can be excited to a higher valance, or orbit, by the energy that comes from photons of the sun. Some materials would require more energy to facilitate this process and for those, sunlight couldn't start the cycle.

## RESIDENTIAL SCALE SOLAR THERMAL TECHNOLOGY OVERVIEW

Solar thermal energy can be harnessed in many ways. The most commonly used solar energy device in Civano, and perhaps the world, is the solar water heater (Figure 7.3 right side). There are several different types of solar water heaters that have been proven throughout the years. Solar water heating systems include storage tanks and solar collectors. There are two types of solar water heating systems: active, which have circulating pumps and controls, and passive, which don't.

Most solar water heaters require a well-insulated storage tank to help keep thermal energy in, especially during night-time hours. Solar-heat storage tanks have an additional outlet and inlets connected to and from the collector. Some designs incorporate dual tanks. Two-tank systems work by allowing the solar water heater to preheat water before it enters the conventional water heater. In one-tank systems, the back-up heater is combined with the solar storage in one tank.

There are three basic commercially available types of solar collectors for residential applications.

## Integrated collector-storage systems

These in-collector storage (ICS or "batch") systems feature one or more black tanks or tubes in an insulated, glazed box. This has been a tried and true process for years, and a favorite amongst "do it yourself" fans. The premise is that cold water passes through the solar collector, which preheats the water. Then the water flows directly to the point of use through a tempering valve or to a conventional water heater, providing a reliable source of hot water. The conventional water heater is considered the backup heater, and the solar heater is considered the primary. This type of solar heater should be installed only in mild-freeze climates because the outdoor pipes could freeze in severely cold weather.

## Evacuated-tube solar collectors

These are an up-and-coming solar water heating technology which is very robust in cold and hot climates alike. They are comprised of parallel rows of transparent glass tubes which are evacuated of air. Each tube has a glass outer tube and a metallic absorber tube attached to a fin. The fin's coating absorbs solar energy but inhibits radiative heat loss.

## Flat plate collectors

Flat-plate collectors are insulated, weatherproofed boxes that contain a dark absorber plate under one or more glass or plastic (polymer) covers (glazing). Unglazed flat-plate collectors are often used for heating pools. They have a dark absorber plate, made of metal or polymer, without a cover or enclosure.

## Active systems

There are two widely utilized types of active solar water heating systems:

1 **Direct circulation systems.** In direct circulation systems, household water is circulated through the collectors and to the home using pumps. These should be used in climates where it rarely freezes or with an automatic drain-down feature that will drain water out of the collector at night so damage from freezing doesn't occur.
2 **Indirect circulation systems.** Indirect circulation systems use a heat-transfer fluid that won't freeze, like an automobile's antifreeze, and pumps that through solar collectors and a heat exchanger in the storage tank. This heats the water that then flows

into the home. They are popular in climates prone to freezing temperatures because they never need to be drained down. Occasionally, the heat-transfer fluid must be replaced, but this is an inexpensive and mechanically simple process that most homeowners would have no problem completing.

## Passive systems

Passive solar water heating systems are typically less expensive than active systems, but they're usually not as efficient. However, passive systems can be more reliable and may last longer. There are two commonly used types of passive systems:

1 **Integral collector-storage passive systems.** These work best in areas where temperatures rarely fall below freezing. They also work well in households with significant daytime and evening hot water needs. The water is stored in the collector and a hard freeze will potentially damage this type of system.
2 **Thermosyphon systems.** Thermosyphon systems are reliable systems which rely on the principals of convection. Water is motivated to flow through the system when warm water rises as cooler water sinks. The collector must be installed below the storage tank so that warm water will rise into the tank. These simple, robust systems are quite heavy so loading must be considered when retrofitting one onto a roof.

The relationships between sun and water effect us all but have been especially important in desert regions for countless generations (see Figure 7.4). Solar power is only as consistent as the sun, which means that there will always be times that solar power is unavailable. To ensure a constant supply of hot water, solar water heating systems are

**Figure 7.4** Artist's rendition of an early Hohokam sun and water petroglyph. *Drawing courtesy of Justin Cupp.*

optimized by the use of a backup system for cloudy days and times of increased demand. Conventional storage water heaters usually provide backup and may already be part of the solar system package. A backup system may also be part of the solar collector, such as rooftop tanks with thermosyphon systems. Since an integral-collector storage system already stores hot water in addition to collecting solar heat, it may be packaged with a demand (tankless or instantaneous) water heater for backup.

## SOLAR WATER HEATING LESSONS LEARNED

Early Civano, phase I builders tried using solar water heaters to meet the SES requirement for solar energy use in homes. The early collectors installed were in-collector storage (ICS) types. Although these types of collectors are generally robust and work well, there were problems encountered both due to the inexperience of the designers and contractors and the incompatibility of the specific ICS product used with Tucson's water due to less than robust design.

- Many of the water heaters were installed on detached garages, much too far from the point of use to deliver hot water.
- Others were installed with temperature/pressure relief (aka "pop-off") valves. The water inside the in-collector storage unit would reach the temperature relief point and the valve would open, dumping the hot water out on the ground. Then, it would heat another batch. The plumbers quickly learned to use pressure only pop-off valves.
- The copper collectors in the units were made by rolling, not extruding, the copper. This process results in long weld seams down the length of the pipes. Furthermore, different grade copper was soldered together inside the unit. These units were from Florida where the water is not as mineralized and they had no problems. However, in Tucson, most of these ICS units eventually leaked and only a few remain in service. Others were repaired or abandoned. Although ICS solar water heaters are a valid design, the remainder of solar water heaters in Civano are made with extruded copper tubes or are the active, indirect circulation type utilizing an antifreeze heat exchange fluid. These have proven very reliable and effective.

## WALL SYSTEMS

The homes and community buildings in phase I of Civano were built with various wall systems, some of which are more unique and others that are quickly becoming commonplace. The community center was specifically designed to incorporate many different wall systems to be a showcase of building materials. The following wall systems can be found in phase I of Civano:

### Insulated concrete forms

Insulated concrete forms (ICFs) are produced in various formats by different companies. The primary innovation regarding ICFs is the lightweight stay-in-place form, usually made of foam, recycled foam, and concrete, or even composite wooden materials. Once the form is erected, steel reinforcement can be added (rebar) and then concrete is poured inside the form. This wall system integrates the strength, thermal storage qualities, and mass of reinforced concrete with the insulating qualities of foam products. ICFs were not used very widely in Civano, but they are very durable wall systems that offer a high level of efficiency, especially in climates where mass is a beneficial component to achieve energy efficiency.

## Straw bale

Straw bale construction was used in two houses and parts of the community center in phase I of Civano. Straw bales offer several advantages, especially in Tucson's climate. At approximately 25 pounds per cubic foot, straw bales are somewhat massive and store warmth or cool well; but straw bale insulation value is quite good as well. Since part of the functioning of most insulation is dependent on the creation of many isolated air pockets, when the inside of the bales are framed, there is a little higher level of airflow which limits the R-value to approximately R-16 to R-18. However, when the bales are fully plastered on both sides, airflow through the bales is minimized and the R-Values go up to approximately R-28. Bales can be stacked by laymen, making the raising of a straw bale home potentially very inexpensive. Bales also come from a rapidly renewable resource, grasses, which is attractive to environmentalists. Figure 7.5 shows a post and beam straw bale home under construction.

There are a few drawbacks to straw bale construction. Since the bales are wide, the wall of the building uses up what could otherwise be living or yard space in the property. The bales require a large amount of plaster inside and out to be finished and weatherproofed, which eats into the potential savings of using straw in the first place. Also, plumbing should not be located in any exterior straw wall because if the interior of the walls get wet, it is a very long and frustrating process to dry the wall out.

Perceptions about straw bale construction are often skewed to believe these structures are inexpensive, but the costs of a home's walls are not a large percentage of the whole costs. If the builder is inexperienced using straw bales or the real estate is very expensive, the savings from using straw will be quickly reversed. However, a straw bale structure can be supremely efficient and comfortable in a range of climates.

**Figure 7.5** A straw bale home unfinished.

## Rammed earth, pressed earth, and adobe

Rammed earth construction is ideal when the earth on the building site can be used; amended with Portland cement, moistened, and pressed inside of wall forms in layers to create walls. Similarly, pressed earth is cement-amended earthen blocks which, ideally, would be comprised from the materials at the building site and then stacked like standard masonry or adobe bricks. Adobe bricks were traditionally un-amended earthen/clay blocks made by hand and dried in the sun. Tucson building codes now require that structural adobe building materials are amended with concrete for durability and consistency of strength.

The walls of a rammed earth structure are often considered quite attractive, with a layered effect that resembles something like the striations in a layer of sedimentary rock. The thermal properties of rammed earth are similar to concrete, adobe, or other solid masonry materials. The effectiveness of the wall system to keep occupants comfortable is proportional to its thickness. In terms of steady state R-value, these types of materials are approximately R-0.2 per inch of thickness. Thus, a 12-inch thick rammed earth wall is approximately R-2.4, a very low steady state R-value (less than a good double pane window with a thermal break). However, whereas lightweight super-insulated walls are as inactive with the surrounding temperatures as possible, massive walls are dynamic and interact with the temperatures around them in ways that can be desirable, or not.

Tucson's large day-to-night temperature swings give credence to massive wall assemblies, but due to their low steady state R-value, a thickness of 24 inches or more is ideal to truly keep occupants comfortable and provide the energy efficiency gains sought. The thermal qualities of massive structures are very desirable because the temperature of the wall radiates into the living spaces evenly and irrelevant of air temperature. Thus, the air temperature can be a bit warmer in the summer and lower in the winter if the walls are radiating a comfortable temperature, and the occupants will still feel comfortable.

To better understand this effect, imagine sitting near to a camp fire on a cold night. Even though the air being warmed by the fire is moving upwards and away, the fire warms people sitting nearby because of its radiant heat. The air in-between the campers and the fire does not necessarily warm, but the campers' clothes and bodies absorb the radiant heat. Massive structures radiate (or absorb) heat and cool in a similar manner. So if a person is in a room which has an air temperature of 65°, but the walls are radiating at 75° due to a day of being warmed by the sun, the occupant inside will likely be very comfortable.

One major environmental advantage of rammed earth or adobe type wall systems is that they are very low embodied energy systems, comprised mostly of dirt either from the site itself or a nearby quarry. So, even if the building uses more energy over the course of its life, the initial impact of building the structure would have been much, much lower than some other systems which require a much larger resource allocation, shipping, and manufacturing. Also, earthen structures return to the earth well, when it is time.

The beautiful round community center building in the heart of Civano, phase I, pays homage to the long time regional building material, burnt adobe block. This structure has thick walls of double adobe brick with adobe interior walls as well. The result is an expansive, open room with very high ceilings and a major attention grabbing hearth-like structure on the northern most tangent of the circular building. This gaping fireplace, tall enough to walk into, is actually the air supply from a very large cool tower (more will be discussed about cool towers later in the chapter).

## Structural insulated panels (SIPs)

SIPs are amongst the quickest up-and-coming wall systems, and many Civano homes were built with them. These are essentially factory built insulation "sandwiches." The "bread" is oriented strand board (OSB) and the middle is polyurethane foam or styrofoam. The SIPs, therefore, are almost entirely comprised of insulation with very few thermal bridges. They use significantly less dimensional lumber because they carry a large part of the building's roof load and provide sheer support. In addition, the panels are often pre-cut at the factory to the prescribed home design, so that they assemble like a large kit on the building site.

Well installed, the SIP system yields a very tight building envelope with nearly continuous insulation and very little thermal bridging. An experienced crew can frame an SIP building quicker than a conventionally framed structure and since most of the cutting is done at the factory, there is little waste at the job site. The price per square foot for the SIP wall itself is higher than conventionally framed, insulated and sheathed walls, but some builders find the efficiencies gained balance or outweigh the unit cost.

Drawbacks of the SIP system are few, one possibly being the necessity to route channels for utilities in exterior walls. Less of a drawback is the fact that these are lightweight wall systems which do not store heat or cool like massive structures do, but given the excellent steady-state insulation qualities of these walls, mass in the wall system is not necessary. The same effect could be gained by exposing concrete slab in sun spaces. This is a very well conceived wall (and roof) system which may very well become an industry standard.

## Steel framing

Steel framing has benefits and weaknesses. Thermally, steel is a very good conductor and a very poor insulator. A wall assembly with steel framing is, by nature, more difficult to insulate effectively because the steel framing itself can become a bridge for heat and cool to transfer into or out of the building. There are steel framed homes in Civano, Al Nichols Engineering, Inc. (ANE, Inc.) being one of them. The insulation strategy of the ANE, Inc. live/work facility is similar to extremely lightweight, icechest like construction. The steel frame is insulated with fiberglass batting between the members and wrapped in two inches of foam board on the outside. The roof was netted and insulation was blown in 14 inches deep and filled up between the trusses.

Lightweight insulation is about containing the heat or cool inside the living space, rather than storing the thermal energy in massive walls. Steel framed buildings are fairly lightweight structures and lend themselves to lightweight insulation approaches. However, finding ways to create thermal breaks between the inside sheathing, the metal stud or purlin, and the outside sheathing is paramount to making as efficient a steel-framed envelope as possible. The Achilles heel of steel framing is the energy performance of structures that were not painstakingly insulated.

One strong benefit of steel framing is its similarity in dimensioning to standard lumber framing, making some of the techniques between metal and wood framing similar. However, the methods and tools used are different and a builder considering switching processes should still budget for costs associated with training team members and transforming their business processes. The costs of taking on a new medium are justified by the value of the new skills and competencies acquired. For builders, the end product of new technology uptake had better result either in process efficiency, lower overhead, and/or improved sales numbers.

Another benefit of steel framing is that it comes from 100 percent recycled or recyclable materials. It is often advertised that the steel is non-toxic as well. While this claim is not in dispute, the larger view of embodied energy reminds us that there is energy spent in the process of recycling a material. Depending on where that energy came from and how it was used, there may be pollutants or toxins associated with recycling steel. Still, recyclability is the major environmental upside to steel framing because most steel studs are made from post-consumer recycled materials. Thus, ore mined and processed into various other products is eventually recycled into a recyclable steel stud. Once out of the ground, that steel works for society for a very long time, if recycling is completed diligently.

# Recycling is a Community Performance Metric

Jason Laros' interview with Gal Witmer:

**Laros:** I'm more familiar with the energy and water standards which are more or less static. What are these moving targets you are talking about that change at different levels of build-out?

**Witmer:** Yea, there is recycling which had a goal, I'm not remembering off-hand, but we wanted to reach about 20 percent recycling for the community. But at some threshold of build-out it was supposed to go to about 40 percent and once we reach 100 percent build-out Civano is supposed to be reaching a high level of recycling. But this has been a very difficult item to track. What do I have to go on? But I've been fortunate in that the City of Tucson has but the blue barrels out as part of a city-wide thing. So I've tried to work with environmental services if we can get a sense of how much recycling the truck going to Civano is hauling out of there; if they're picking up more recycling there than in other areas. What they told me is that they don't have a means to study that because these trucks are going to places other than Civano. They're going to too many places that they can't isolate Civano's recycling. So there are things that are set up so we can't really track easily.

Also, when the homes are built, the IMPACT standards said that the builders should use some recycled material. But it wasn't clear if this means that they should recycle some building waste, or if it is about using materials with recycled content. There was also no quantitative amount. It just said "recycled content" so, you know, when Swaim started trying to enforce these things we got push back. The builder would say, "We do, OSB has recycled content." And you really couldn't argue, as long as they could list on a piece of paper that they had OSB, pink batt, or fly-ash. So this is one of the real issues I see with it.

I suppose that if Pulte wanted to, they could really take more of a stance on these things and revise it to create some harder goals, but you'd have to have the support of the people to track these things.

**Laros:** Absolutely. Pulte has picked up the Civano project and kept it alive. They are providing several very excellent housing products in Sierra Morado, but when I go over there and see what goes into their trash dumpsters, it is occasionally disheartening.

**Witmer:** You know, they could be using waste management like they do for LEED certified projects. As long as they had bins labeled and tracked everything, they could be recycling a number of items. There are just not enough teeth in the IMPACT language … as long as they say there is fly-ash in the concrete, well, they've got the standard.

But you have to realize, too, that this was all written 15 or 20 years ago when nobody was even doing this or asking for it.

Sections from: **Impact System Monitoring Report #13**
**3.4.6.4 and 3.4.6.5 Reduction of Wastes**

As part of the review process, buildings are required to provide built-in recyclable separation features. Residentially, this is often accommodated with recycling bins incorporated into the kitchen cabinetry. Commercially, both interior bins and a screened area are often provided for multiple recycling containers. Both the neighborhood center complex and the Civano Community School successfully utilize only residential style trash and recycling roll-offs in lieu of the dumpsters typically seen in Tucson. In 2002, the City of Tucson introduced curbside recycling—the blue barrels. This move significantly helped the Community of Civano meet this goal. Per the City of Tucson's recycling Web site the current city-wide rate is 21 percent. The performance target of the IMPACT standard for Civano and Sierra Morado at outset is 30 percent increasing to 60 percent upon 1250 residential units being constructed. Although an on-site recycling center has not been established, the community garden in Civano 1 does utilize composting.

Builders and designers are encouraged to utilize green and other recycled products in their projects and are required to list the utilized products and methods during the design review process. As far as the limiting of construction waste, Pulte and other builders typically work with their sub-contractors to identify areas of waste and coordinate processes to get recyclable materials into the appropriate places. Pulte has also looked at moving to pre-fabrication techniques and less options on models to eliminate over-orders and areas for error leading to wasted materials.

Establishing quantitative amounts for the recycled materials utilized in homes and the recycling of waste would be a next step in increasing the performance targets. Drawing from Scottsdale, Arizona's "Green Building Program Guidelines," it is a pre-requisite that a built-in recycling center with two or more 5 gallon bins be installed in or near the kitchen. Also, points are to be gained by builders for "Construction waste reduction/reuse plan written and followed (e.g., recycle bins for wood, cardboard, drywall, foam, metal, concrete, masonry, asphalt): 1 pt. for each item recycled by builder or waste hauler." Other point driven choices: "Wallboard made with min. 25 percent recycled material (e.g., wheat board, drywall with flue-gas or industrial by-product gypsum)": "Paints/finishes with minimum 20 percent recycled content"; and "Reconstituted or recycled-content fascia, soffit, or trim, (minimum 50 percent pre- or post-consumer), OSB, or w.p. gypsum board. The City of Tucson's Environmental Services department can provide builders different bins to allow job site recycling now. This will be required in order to calculate fully if the community is progressing toward meeting its goals.

Although every effort is made to reach target goals at Civano, there have been challenges, as of yet, unmet. Brian McDonald discussed some of what could have worked better at Civano with Councilwoman Shirley Scott.

**McDonald:** Could you elaborate on what some of the lessons learned at Civano were?

**Scott:** There were certain appliances that were tried in Civano that didn't work properly. Over time they were replaced by those which did. Part of the original Civano plan was to require an electric outlet in the garage to recharge electric vehicles. As a matter of practicality, there are very few electric vehicles in use, and the insistence in the code that you put that electrical outlet there became onerous, since no one was using an electric car. That requirement has since been deleted from the code.

For the most part, Civano uses standard technologies to achieve high-performance goals. This leads to a robust and more easily maintained stock of housing than if they were each custom or highly exotic. There were devices and materials, such as heat pump water heaters and single coat stucco over OSB, which simply were not great ideas. The heat pump water heater was a complicated device that was not well supported. Breakage was usually followed by replacement rather than repair. One contractor thought that a single coat of stucco over SIP panels (OSB) would adequately weatherproof and finish a home. It did not. They had to do that work over again. Many small, fussy details live in the various components of a building. From an engineering standpoint, those approaches and products that have been tried and were proven to work well are already ahead of the game.

## HEATING, VENTILATION AND AIR CONDITIONING (HVAC) IN CIVANO

With few exceptions, Civano homes are heated and cooled by high efficiency air-to-air heat pumps, or high efficiency air conditioners and gas furnaces. There are a few homes that utilize evaporative cooling a great part of the year and even one custom home which has a cool tower. A cool tower is a device akin to evaporative coolers that are found cooling buildings in many arid environments. However, instead of using the power of an electric fan to motivate air past water-laden pads, the cool tower captures breezes from outside and channels them across pads. As the heavier, water-cooled air enters the large tower, it falls down the shaft and creates some small amount of suction which drags more air into the tower and across the pads.

Cool towers are based on very old technology. Modern cool towers only need a small water pump. They do not need a fan motor and they can be fairly effective if all conditions are exactly correct in terms of wind, humidity level, and tower design. In comparison to a modern evaporative ("swamp") cooler, these gravity-driven devices are very elegant, but cool towers are not the epitome of green design. A cool tower consumes many more materials, uses water less efficiently, and is often much less effective than its mechanized swamp cooler counterparts. In fact, some swamp coolers are available now which can run on as little as 50 watts, or a single 85 watt photovoltaic panel. There are many more exotic ways to mechanically condition spaces in a building, but the refrigerant and compressor combination is a tried-and-true method of creating a temperature differential useful for cooling or heating air.

In general, Civano has stayed at least one seasonal energy efficiency rating (SEER) number ahead of the local energy code requirement. The SEER number indicates the efficiency a specific unit will achieve in a specific climate if it is properly sized. So when the international code required SEER 10 units, Civano builders generally installed widely available SEER 12 units. At the time of this writing, SEER 13 units are the minimum efficiency being installed in buildings at large, so SEER 14 units are being used in Civano homes.

The ANE, Inc. live and work facility uses heat pumps for both the office and living quarters, and an evaporative cooler for the living quarters. Most heat pumps are specified to include backup electrical resistance strip heaters to help with wintertime heating because in very cold weather heat pump coils tend to frost over, go into defrost mode, and stop being able to provide heat. When the heat pump cannot maintain temperature, the (inefficient) electrical strip heaters fill the gap. ANE, Inc.'s heat pumps do not have electric back up heating because in Tucson the heat pumps may only struggle to provide adequate heating a couple days per year. Many people choose to turn off their backup heat strips (at the house circuit breaker) for the same reason—to save electricity.

With standard equipment efficiencies increasing, it is somewhat difficult to justify the premium associated with more efficient (and complex) mechanical systems. The minimum efficiency gas furnaces being installed in the village are over 90 percent efficient at capturing the heat energy of the combusting gas and air. These units burn very cleanly and use a sealed combustion air cycle that takes cold air from outside, combusts it with the gas fuel in the furnace, and vents the gasses directly back outside. The heater never sucks combustion air in from the house which helps keep the warm air in the home.

In some ways, gas thermal heat is a more efficient use of energy than electric thermal heat. The cost of delivering the electricity energy through the grid is nearly 66 percent. That is, for every 3.1 kilowatt hours produced at the plant, approximately 1 kilowatt hour gets used at the house. Then the use of a compressor or strip heat has further losses in efficiency. Natural gas takes only about 10 percent of its energy to be delivered to the home and efficient furnaces only lose approximately 10 percent of the fuel energy; meaning gas is more efficient to deliver and more efficient to use for thermal heat. Still, with the costs of gas on the rise and a perception of higher market volatility, many people still choose to purchase all electric homes.

## Water and air

By intuition, one knows which is more massive, water or air. Which weighs more, a bucket of air or a bucket of water? Just like massive wall systems versus lightweight super-insulated ones, water holds thermal energy and reacts with its surroundings dynamically whereas air is less affected. There are questions one will naturally ask about the thermal capacitance of water in different physical states—as vapor, steam, or ice. Sometimes, water and air seem indistinguishable, as in a humid climate where one breathes the air, yet is aware of the moisture, even when it cannot be seen. The relationships between water and air help make our dynamic earthly climate possible.

Water can conduct much more thermal energy in a much smaller volume than dry air. An enclosed column of water in plumbing is not very compressible or expandable, but it does move fluidly. Air, on the other hand, is very "stretchy" in the sense that it can be compressed or expanded easily, even inside an enclosed space like ductwork. Furthermore, air's humidity and temperature qualities change as the air does.

If moist air is cooled, water condenses out, just like rain from clouds. If air is heated up and it expands, it will absorb water (if any is present) until the air is saturated for its given heat and density. Air has its limits about how much thermal energy it can carry at a given density and water content. There are limits on how far it can expand in a given space and maintain temperature, or how much it can be compressed at a given temperature before the water comes out of it.

Given its properties, water is a very attractive and efficient way to transport thermal energy to where it is convenient to use because the piping required to carry the same amount of thermal energy in water is much smaller than the size of ductwork needed to carry the equivalent thermal energy in air. Moving large volumes of air required to transport thermal energy with fans is an energy consumptive process. While pumping water can generally be very energy consumptive, it takes relatively little energy to pump water around in a loop of piping and heat exchangers which have been well designed. The trick to efficient heat transfer is passing the water or air (transport media) through a medium that offers high conductance between the hot or cold source and the transport media.

ANE, Inc. often receives queries about ground source heat pumps. These are units that send water through loops of piping in the ground (or a body of water) to make the water temperature closer to the ground temperature. Then the tempered water can be further heated or cooled using mechanical means. Or the water can be used to temper air which then becomes further heated or cooled using mechanical means. Either way, the concept allows for efficiency gains because mechanical work is replaced by geothermal energy, which is abundant and essentially free.

Ground source heat pumps are used in many parts of the country where conditions are favorable. Ground source heat pumps are superior in efficiency to air source heat pumps because of the various thermal mass qualities of water and air. So in places like Florida, where the ground is soaking wet, a conductive pipe looping underground will conduct ground temperature from the highly conductive wet soil into the highly conductive water inside the pipe. Although the ground temperature is favorable for ground source heat pump use in the Tucson region, unfortunately the soil is very dry and has poor thermal conductance.

There are a few ground source heat pumps installed in these conditions, but given the same ground temperatures, it takes approximately 30 percent more piping to achieve the same water temperature differential in poor soil than in ideal soil conditions. Therefore, in some places the added costs of ground source heat pumps are even greater to achieve performance targets. If these conditions are true, the advantageous qualities of water for transporting thermal energy cannot always be effectively utilized when compared to the most viable alternative.

The relationships between air and water are very dynamic and must be understood in detail by HVAC engineers. If they are not aware of the air humidity and temperature qualities in a given space, they may unwittingly create situations where warmer, moist air is in the same space as cold pipes or ductwork with surface temperatures below the dew point of the air. If this happens, water condenses onto these colder surfaces, like droplets on the side of a glass of ice, and things inside the building start to get wet. Water where it is not supposed to be can lead to anything from cosmetic and structural damage to sick building syndrome.

Sick buildings are often associated with mold growth due to the improper management of air conditioning or plumbing systems. With bad luck, indoor mold could become an allergen for people living or working in the building. With worse luck, the mold spores which take hold could be one of the few kinds which are toxic to everybody like the infamous stachybotrys genus—the "black mold." There are thousands of different types of molds, many of which are black in color. There are only very few types which create the mycotoxins stachybotrys is known for and thus, very few types of mold which are directly poisonous for humans and animals and a relatively low likelihood that any given mold colony will be toxic. However, in settings such as hospitals, any mold is considered highly dangerous because individuals with decreased immune functions can be susceptible to grave illness from even non-toxic molds.

## Passive structures

The discussions about mechanical conditioning of spaces can carry on ad-infinitum. The ideal structures would consume essentially no energy to heat and cool themselves. It is as if the mechanical systems are integrated into the home. This type of building requires a large amount of early design effort, but in many climates, natural ventilation can provide a large amount of air circulation in a building or home. Convection can be capitalized on by the well-planed placement of low and high operable vents and/or windows combined with either the acceptance or rejection of sunlight into the space. Massive structures such as masonry walls or concrete slabs or even bodies of water can be utilized to absorb or release heat or cool.

One stalwart example of such a structure is the Rocky Mountain Institute (RMI) headquarters in Old Snowmass, Colorado. The structure was completed in 1984 and designed with the idea of collecting enough solar thermal energy that mechanical space heating would not be necessary. The structure was built with double the insulation required by the building code of that time with a very tight envelope and very advanced windows designed to let light in and trap warmth in the space. This passive solar designed building is reported to save 99 percent of the water and space heating requirements that would otherwise be necessary for a 4000 sq. ft. structure. Unlike in Old Snowmass, Tucson's climate (and other very hot climates) makes heating not as much of an issue as staying cool.

The problem of staying cool in a very hot, sunny environment can be more of a challenge to do passively than heating is in a cold, sunny environment. Large benefits are gained by protecting windows from direct sunlight by using shade structures such as awnings, fins, or trees. However, if the nighttime temperature never drops below a comfortable temperature, even opening all the windows and doors at night will not bring the building temperature down enough.

There are many different passive approaches to staying cool in the desert, ranging from using underground ductwork to cool ventilation air as it is channeled into the space, to building partially underground or earth-berm structures. Some people even plant water-thirsty vegetation all around their homes to provide shade and a slightly cooler microclimate, but even if that is partially successful, the environmental costs of providing the landscaping water often outweigh the incremental benefits of trying to change the microclimate around the structure. In the end, some conditions cannot be met with fully passive systems. There needs to be the addition of water to air, or even mechanical systems to achieve greater levels of comfort.

The idea of Civano is to help achieve replicable results on good outcomes. This goal requires that the systems used are not too obscure because even if they are very effective, it may be difficult for builders to adopt very different practices and keep costs reasonable. Providers of new services or goods might or might not be around in a few years to honor warrantees. New approaches simply have start up costs associated with them and these costs will usually trickle down to the customer. Therefore, innovations should be as aggressive as possible without overstepping the tolerances of what can be sold, maintained, and supported.

Many of the most practical ways to increase the performance of buildings, and most of the innovations of Civano buildings, are items that are not readily seen. So, what does building science innovation really look like?

**McDonald:** What do you know about the building practices out there? There're two straw bales, one rammed earth, and one other nonconventional home out there. And as far as I can tell, you know, there's a burnt adobe brick out there. And when you look at the buildings out there, it looks like conventional, you know.

**Singleton:** It looks rather conventional, but they have—all of them have heavier insulation in the walls and in the roof; whereas the energy code would only call for like 2 x 4s, stick stucco would only call for an R-11 in the walls. They went with 2 x 6s and went with R-19 in the walls. And whereas the code would—ICC would only require R-30 in the roof, back then they were going with R-38. And I think some of them even went higher than that, R-40s in the roof. Plus they used evaporative cooling during the times of years when it's effective and a water harvesting area plus the solar panels to heat water.

All of that and some of them are using photovoltaics, as I understand it. And all of them have to use a certain percentage of alternative energy sources. And to my knowledge a lot of them were using a lot more than the 5 percent.

Other than a unique neighborhood design incorporating new urbanism features, the visual look of most of the homes in Civano is fairly typical. Stucco and block finishes with parapet, tile, and metal roofs. Even high performance doors and windows look somewhat like any other doors and windows and tell little about the efficiency of a building to the typical onlooker. The only obvious feature that stands out in Civano are the hot water solar panels that are found on nearly every home, and the photovoltaic panels which are becoming ever more popular. Other than that, it is pretty difficult to discern the qualities that make Civano a more sustainable community. When asked if Civano was a success, this is what Bob Cook said:

**Cook:** The performance goals are all positive. I think that the next phase is to democratize this. And to have some of the benefits of the Civano vision be applied to entry-level houses in lower income developments so that people at all income levels can begin to actually buy into some of the benefits—that is a measure of success. Energy savings is something that really ought to be affordable and should be accessible to everyone. And if it's marketed right, it could.

# 8

# A MIDDLE GROUND—PHASE II

## The Costs and Benefits of Production Housing

*A good compromise, a good piece of legislation, is like a good sentence; or a good piece of music. Everybody can recognize it. They say, "Huh. It works. It makes sense."*

—BARACK OBAMA

The failures of Civano are not this story. They are the moral of the story. What make this story are Civano's successes. The urban design successes of neighborhood I overshadow much else about the project. It is such a pleasing place. It is a place. To many people, only this one part of the master plan is Civano. The rest of the community is separate. The first phase community has roads, paths, and trails that are like arteries, veins, and capillaries where humans flow. The other phases do too, but some would say they lack the heart of neighborhood I.

As one walks through neighborhood I, many scales of roads and buildings set a continuum of change from the lower slung detached single family dwellings into all two story, mixed use row-type housing toward the town center. It feels natural to see the tall cool-tower that signals arrival to the community flagship, the community center. Along the way, shaded by mesquite and serenaded by birds, it is an easy place to walk, bike, or slowly drive. Of course, the proverbial "arm and a leg" are in there too. The place was expensive to build. The mega equity investor Fannie Mae never fully adapted to the Civano project because their primary role is not as a property owner or manager. They eventually sought a large developer who could finish building out the remainder of the project.

**Lee Rayburn:** I tried to convince Fannie Mae that we could all have our cake and eat it too. Pulte is a great builder. They understand energy conservation. If we could just help them understand that a neo-traditional plan could be done in such a way that they don't have to reinvent what they do, we could create the look and feel of Civano neighborhood I that has been successful; and we could maintain the energy standards without wholesale diminution of their goals and impacts. We weren't allowed to have a classic new urbanist charrette, but we did have public workshops. In these, we used basic new urbanist planning concepts to show how the future neighborhoods of Civano could look.

**Figure 8.1** The main entrance to Pulte Home's Sierra Morado at Civano (phase II).

That was when Fannie Mae had to make a decision whether they could step out of their comfort zone a little bit to adhere to the unique master plan begun with Civano neighborhood I, or let Pulte develop a master plan that was more consistent with their comfort zone; which was the tried, the true, and unfortunately mostly devoid of the special aspects of Civano neighborhood I.

Expense is what balances the equation between the first phase of Civano and Pulte Home's Sierra Morado (Figure 8.1). Although high expense does not necessarily equate to advanced urban design, typically lower cost housing means more traditional urban and architectural design—or a lack thereof. Be that as it may, Pulte's work as master developer of the largest portion of Civano has come under both scrutiny and praise. The latter is probably much more appropriate given the fact that award-winning Pulte Homes never needed Civano as much as Civano needed Pulte—and Pulte delivered many affordable housing units which meet the Sustainable Energy Standard.

By now the reader understands that this has been at the crux of the Civano story moral—it started off too big, so big that less flexible giants like Fannie Mae and Pulte Homes were the firms that came to finish up what was started by a smaller, more specialized company. The ideas, although very visionary and progressive, may have been too big as well. There were ultimately compromises made on behalf of the project stakeholders and end users and some of them are still upset to this day. However, the top-down integration and efficiency of a builder like Pulte brought SES-quality homes to the masses at an affordable price, a full decade or more before homes of equal energy and water performance quality would be mandatory by code. And they included solar water heaters on every home to boot. Perhaps Pulte knows how to balance the scale. Or perhaps they don't know how far they could have pushed the boundaries of production housing.

**Rayburn:** I knew Fannie Mae had to have a national builder. Of all the national builders, no question Pulte was the best choice for this project. So I was a fan of Pulte, but here's the difference. I firmly believe that the success of Civano neighborhood I is about 70 percent look and feel and the rest is all the other great benefits. And the longer you're there, the more incredible you realize it is. For me, and this is part of new urbanism, when you go down to Sierra Morado, I don't think you are going to see people walking or riding a bike down there or kids out playing like you do in phase one.

The discussion about which is the heart of Civano, the new urbanism or the engineering, is a root philosophical debate which helps us define what makes a great master planned, high-performance development. The differing landscape and urban designs of the adjacent phases of construction offer great insight into the world of environmental perception in our built environment. It is a place for questions like what makes some people express distaste for the southern two thirds of the project after seeing the first neighborhood? What did Pulte Homes have to discard from its design to make its business model satisfactory? Which urban design strategies would have added the most value if they were left in, and why? Perhaps most importantly, can mass production techniques be adapted to include new urbanism while achieving the same economies of scale that have become ubiquitous with affordable housing? In essence, the goal is to increase quality of life as much as possible while increasing cost as little as possible—or not at all.

Typically in mass production, each unit sold may yield less profit for the builder or developer, but they build and sell many units quickly to make up the difference. Less profit per unit also means a smaller margin of error before a particular unit or project changes from an income-producing asset to a money-losing liability. Thus, it becomes necessary to know exactly what a builder can technically do and how fast. In Civano, Pulte built standard designs using standard wood framing techniques. They included enough envelope and mechanical upgrades to meet the energy codes and enough water saving features to meet the potable water requirements.

Pulte continually rolled new product configurations off the shelf as the build-out continued, but the overall architectural style, fit, and finish across the board are all similar. This homogony of Sierra Morado is one of the most commonly stated observations about the stark perceptual differences between the first neighborhood and Sierra Morado. Sameness between units saves cost because processes can be more easily streamlined as personnel get to know their moves and the materials they are working with. The actual opportunity cost of making the decision to make very similar homes across phase II of Civano was the kind of sense of place found in neighborhood I. Although sense of place is difficult to quantify in financial terms, this is exactly what new urbanism scholars and professionals are leading the industry to do.

The challenge is determining which qualities of the new urbanism design tools give a place a solid identity and which ones are arbitrary. Sierra Morado has somewhat curvy roads and a couple large park areas that one can key into, but there is no well-defined center or other specific landmark to relate to. Almost every building is two levels with only sections of the buildings, such as garage roofs, at single story height. Ultimately, there is a feeling of sameness in Sierra Morado no matter where one is. Each and every roof has the same kind of tile. Each home has the same basic stucco finish with only small accents and variations. The detailing of each home is very similar and the color palate is limited to a few safe, earth tone colors. This makes it easier to get disoriented and enhances the similitude of architectural effects in the community.

The very way one speaks about some very commoditized housing developments is evocative of the culture that created them. The homes are "units" to be sold. Each unit is from a different product line and each product is designed to appeal to a specific cohort of consumer. Overlay a mass-production ethic on top and the urge is to cram as many units onto a piece of land as possible, chosen from the standing product lines in a mix that is expected to sell well to the local mix of potential home buyers. Some housing developments are almost like housing product warehouses; the way homes are stacked in rows and waiting to be sold. But the "warehouse" is actually the neighborhood where people will live.

Just as Henry Ford helped bring the Model T to the masses using mass production techniques, what is gained from standard production housing are process efficiency and ultimately money. The result is generally a more affordable housing for consumers. What gets lost, and what many people feel has been lost from Sierra Morado (when compared to the first neighborhood) is a sense of place or community—and with it, a degree of function. Assuming that Pulte pushed its business model as far as possible in negotiating the Civano development agreement, which design elements did Pulte Homes have to discard from the design to make its business model satisfactory?

# The Community Center

Pulte did build a community center. The 13th Impact System Monitoring Report by Gal Witmer states:

> In Sierra Morado, the community recreation center is complete, providing a pool, tot lot, basketball courts, meeting rooms, demonstration garden, event lawn, and promenade. It has taken a different approach than Civano 1 by providing only one central large community and recreation center as opposed to two smaller centers with limited parking dispersed through the community. The Sierra Morado center provides a large parking lot and its distance for some in the community may encourage automobile use to utilize it. However, this will be a personal choice as many American's are reconsidering the use of gasoline as the price continues to soar. (3.4.6.6 Reduction of Auto Travel)

The Sierra Morado community center is an asset to the community. As urban housing densities increase, lot sizes decrease. In places like Civano, what gets displaced from lots are mostly yards. The tradeoff for yards is shared community space. It stands to reason that the quality and usefulness of the shared community spaces must in some form encompass or surpass the value of the yards they replace. The difference is that people have to come out off their lots to enjoy some of the amenities that used to be found in backyards. In many cases, however, the community enjoys benefits of amenities which could never fit in a small suburban yard.

This line of thought may lead one to consider that the draw for somebody to leave their lot is a part of the value provided by community resources. For some people, it may be for the love of an outdoor barbeque, and for others it may be solitude in an outdoor, natural setting, or a swim in a full sized pool. These things can be provided in some form by the spaces in-between the homes and the amenities.

In Civano's first neighborhood, the involvement of the community center shows considerable thought about the "places between." The primary difference between Sierra Morado and neighborhood I is that Sierra Morado lacks a true focus or center which serves as a major anchor point. One can find several different ways to traverse from home to other parts of the community. In some cases, the trails which serve this purpose are one of the very amenities which provide a replacement use for a yard. For example, the desert country trails in neighborhood I provide a recreational, nature-oriented use and also serve the function of being a transportation corridor for pedestrians walking to work, the pool, school, park, or the neighborhood center district. In Sierra Morado, there are sidewalks, bicycle paths, and streets that can support most modes of transport. There are parks and even small areas of centralized parking for residences. But there is no particular draw to any particular part of the community.

There were other less significant features of Sierra Morado which are clearly not with some of the mainstream thinking about neo-traditional or new urbanism-based design, such as the use of "front-loaded" products, or homes with the garage in front. These bulbous garages, in some circles, are considered the snout of a front-loaded product and Pulte Homes used many of these models, especially in the further reaches of the neighborhood. Nobody can say for certain if those furthest out homes will ever be any closer to amenities than those provided within Civano itself. This lends some credence to the greater emphasis on the automobile infrastructure in the homes only marginally within reasonable walking range of the nearest amenities. There may have been other ways to integrate those homes into a more dense urbane format than the suburban feel they have; but the current use is logical from a production housing standpoint.

In other places, Sierra Morado evokes row housing, some of the most affordable units in Sierra Morado (at the time of this writing) have their fronts turned toward one another and their garages serviced by short alleyways. Visitors can put their cars in small parking areas and walk from the blacktop to a sidewalk which winds in-between front doors. In other lots, garages of four units share a single concrete driveway. Many of these houses do not have any yard at all, nary room for a small barbeque on the front porch.

There was community interaction with Pulte during the planning, design, and building of Sierra Morado. Parts of this process are recounted by Simmons Buntin.

**Buntin:** My name is Simmons Buntin. I first heard about Civano in about 1995. I worked for the U.S. Department of Energy, for an agency called Western Area Power Administration. I was an energy services program manager. When we heard about Civano, we were intrigued. It wasn't within one of our customer's specific areas but it was something we were promoting both from a renewable energy and demand-side management perspective.

So I heard about it through that and then I heard about the new urbanism concept and that really intrigued me. When I was in graduate school, I was at the University of Colorado in Denver, and got a graduate degree in urban and regional planning. Graduated in 1997. I lived in Tucson as a kid, and my wife grew up in Phoenix. So we decided to move down here. I had a couple opportunities to come down and visit Civano. My agency held a meeting down here and I came down here when construction had just begun. There wasn't too much on the ground at all. I just kind of swung through here and got some of the original information.

Then we had another conference down here and we came back down in 1999. It was after the neighborhood center was built and opened. And we were convinced we could move down here. So I started looking for a job, found a job, and came down. We lived up near Sabino Canyon for eight months and then put a contract on a KE&G home, which we still live in now. At the time we really wanted an RGC desert country home, but right when we started calling is when they weren't answering the phones and fortunately for us, we didn't get stuck in a contract that turned bad like it did for others. We've been really pleased with our KE&G home on Sixto Molina Lane.

**Laros:** You interface with the community in various ways. How did you first get involved?

**Buntin:** When we first moved here, there were potlucks in the neighborhood center and we came out for that. This was before our house was actually built, or even under construction. We met a lot of people out here, probably Al I imagine. One of the community members was there and told me that with my technical background they'd like me on this committee that was going to explore how to further integrate technology into Civano, such as wireless internet.

I agreed to head that up. In turn, I started to get involved with neighborhood politics, and hanging out with that group. Our house was finished in November 2000, one of the very first ones on the street of KE&G Homes. So that's how I got involved, it was this telecommunications committee. And we had a series of meetings to talk with Sprint about bringing in wireless internet. We were talking with Lee Rayburn about that.

Anyway, a little while after that, there was a discussion forming about expanding the existing website, providing a little more information. We were talking about almost doing an intranet, or an extranet from the community perspective. So we started the Civano Neighbors website. I can't remember if the whole thing may have initially been password protected. But that was kind of tugging along there.

I guess I got much more politically involved when some neighbors realized Civano wasn't quite building out like we wanted it to. I remember what specifically made us angry was what was going on in the Desert Country area. The developer was putting down a bunch of rip-rap where the trails were supposed to be. They were putting in a bunch of big rock. Bednar was building there, before Pepper-Viner came in, and a lot of Bednar homes seemed to be facing the alleys and treating the alleys as streets, really backwards. Then Contravest came in and built some of their models and they clearly faced them toward the alleys and not toward the walking paths. We were concerned that the houses were not sited correctly, or at least the way that was intended by the specific plan.

We tried to meet with Fannie Mae, and they refused to meet with us. So, a bunch of neighbors got together in Rick Hanson's house to decide what to do and we decided to write a letter that demanded they stop construction in the Desert Country area until this issue got resolved. I was the guy who drafted that letter, which said in 30 days we expected to be in discussion about this topic, but nothing happened. When nothing happened and 30 days had passed, we notified the city about our concerns, and the city came out and was likewise frustrated. This eventually led to the formation of the neighborhood association. We decided we needed to have a larger voice about what's happening at Civano than just the neighbors on the advisory council.

**Nichols:** If we had maintained our leadership and it had not been turned over to the banks and large developers, would the original developers have been more successful?

**Buntin:** Yes. You know, the new urbanism communities, in my experience having a degree in urban and regional planning and studying new urbanist projects, you have to have a developer who cares about it, who stays in it for the long run, is consistent, and pays attention to details. Obviously, when Fannie Mae came in we really didn't have that here. There's all kinds of calculations about what was going on with Fannie Mae or not, but the bottom line is that they didn't really care and they didn't pay attention. They just wanted to build out and get out, I think.

I think that if Kevin Kelly and David Case would have stayed on, Civano would have built out differently. It's weird, though. On the other hand, a lot of new urbanist projects stall from a commercial perspective. But I haven't heard of too many that had the kind of challenges Civano has had. Part of that was the black eye of RGC coming in and then pulling out on all those contracts and part of that was the additional black eye from Contravest when it went under. But that was citywide, not necessarily specific to Civano. I know that people were frustrated because the same time Civano was building out slowly, Mesquite Ranch built out really quickly. But it was a more traditional subdivision model. So perhaps Civano was too high-end initially for most of the Tucson market. Of course a lot of the people I know here came from out of the state, they came from all over to be here.

**Nichols:** There is something you said, which I have already said, is that the project was too expensive for our market here. Phase I just became expensive. There were too many fingers in the fire.

**Buntin:** Yeah, I'm not exactly sure why it was expensive. Coming down from Denver where the price of housing was pretty high, the cost for phase I seemed pretty good. I mean, it was high compared to the rest of Tucson; a house here was probably 50 or 60 thousand more. I mean, we have more amenities.

**Nichols:** We're about 15 percent over market. And we've been saying that the architecture was about 10 percent and the energy standard was about 5 percent. We've been comparing the first Civano neighborhood to Sierra Morado. Then the whole issue becomes, "Are the later phases of Civano meeting the goals of the project?" I think that our inner consensus is that we lacked a plan book. We had one, or we almost had one, but it didn't get truly adopted, so that was an upfront error. I don't know where that would have come in, or where it should have come in.

**Buntin:** Well, it should have come in about the time the specific plan came in. It should be part and parcel with that. Here, the specific plan has broader context. Here's where the streets are going to be, here are the land uses you can have on that. The pattern book is streetscape level. Here is what it can look like, here are the options, here is what can and can't go next to each other. The specific plan does some of that. It has the house on the lot and the spacing requirement and the like. The pattern book would have taken out many of the unknowns.

**Nichols:** Truly, the streetscape defines the walking community.

**Buntin:** Sure it does, because that is what you see when you walk.

**Nichols:** And that is what defines new urbanism? Walking paths and walking communities?

**Buntin:** Well, that is a lot of what it is about. But really, new urbanism isn't just the street level. It is even about the whole regional level. Streets, in an urbanist community, should be multi-modal. They should work for the pedestrians and the automobile. So when you put cul-de-sacs in, that's really only working for one. That's really only working for a person. It's not working for a car because a car can't get through. Only, this is a mindset. Pulte didn't want to or choose to care about that or make it a priority.

**Nichols:** Why?

**Buntin:** Because the city didn't make them do it.

As a neighborhood group, we had that year-long conversation with Pulte and it seemed like we were handed from one management team to another management team and we basically told them they were backing out of a lot of promises. At first it was no more than a very traditional suburban subdivision lay out. But we were a neighborhood group, with no authority. We were lucky that Fannie Mae ever told Pulte they needed to get the neighborhood buy off on their work and get our participation. But then the city let Pulte off the hook in many ways. You know, letting them change the definition of what front-loaded versus rear-loaded is and saying that's okay.

**Nichols:** Why do you think that happened?

**Buntin:** I would say it is just developer driven. I mean, that is the mindset. We've got to build, we've got to grow. We have to keep going. And that just is the way Tucson is. And that was this great gem of Civano neighborhood I, that you could do things differently. We had all these challenges which may have given it a bit of a negative reputation. So it was much easier for the city to look at the traditional land use code even though there was a specific plan and even though there was the master plan that they should have been judging the development in Civano against. I think they felt like they just wanted to get it done and move on.

Politically it may not have been easy, but I think they could have held the line. There would have been a lot of pressure from Pulte, but the right thing would have been to hold the line from a land use perspective. They were unwilling to develop a pattern book or even do a design charette like we had done for Civano. To a lot of new urbanists those are critical if you're going to do something. They were really unwilling to be specific.

With criticism come the facts that Pulte did meet the Civano open space requirements. Some of the open space is comprised of very functional park areas that double as detention basins. One notices that the community undulates with more natural topography than the first neighborhood. This is because Pulte worked with the site's natural hydrology more than the first neighborhood which was built on a fully engineered sight—requiring the blading and grading of approximately one to three feet of dirt across the entire first phase to accommodate their ideal road lay out and urban designs. The undulation in Sierra Morado adds a natural dimension that helps take the edge off the identicalness of other aspects. After exploring there for awhile, one picks up on the hills and valleys in the landscape and they become location cues - part of the sense of place.

Al Nichols and Brian McDonald discuss Pulte's branding efforts at Civano:

**McDonald:** Do you think that Pulte kind of pulled away from the name Civano?

**Nichols:** They wanted to pull away; oh, they did. They wanted to not be required to use the solar, which was the first thing. They wanted to line the houses all up like barracks, which even I thought was disgusting.

**McDonald:** But I thought that they did.

**Nichols:** No. From their first presentation to what they actually built is quite a bit different. The new urbanists kept beating them up to design more curvy roads, more colorful landscaping, and other aesthetic stuff. So they did, to a degree. But they didn't have any California architects coming over. They didn't work with any new urbanists. They engineered their homes. In fact, when I was on the Metropolitan Energy Commission, we sent them a congratulatory letter for getting a national energy award for their homes. And they build thousands every year.

So they met our standards, okay. We did let them off on the reclaimed water. Reclaimed water turned out to be a disaster when delivered on too fine a scale, to every home. It is expensive to have the infrastructure in the first place, and expensive to keep it, and it's expensive to buy the water. It cost about 25 cents a ccf (748 gallons) more than potable water. You had to spend $1600 to have the meter. You had to pay the meter charge. And you have to have a backflow preventer which, in turn, has to be tested every year. Reclaimed water also has so much mineral in it which results in clients having a difficult time with it clogging emitters and building scale. Since reclaimed water is usually used for larger-scale irrigation, it was delivered at too high a pressure—it was delivered 100 psi—for residential irrigation systems to handle. So it was just bad news, bad news, bad news, bad news—but that is why we have the IMPACT process. We were able to adapt and make sure that the developer didn't get stuck with a bad deal. They didn't back away from the energy or water standards though, which had already been proven highly feasible.

So, Pulte has the end game of Civano, filling out the rest of the project. For all the developments they've built over years and years, and real estate they've sold, they have had to decide how this project is different and if it has an advantage in the marketplace going forward.

Watching Pulte gives us an idea of how a large builder makes decisions. They have had experience with master planned communities and they have built energy efficient homes before Civano. But the strengths they exerted on Civano were oriented toward building homes and less toward building a community. They did easily adapt to manage the mechanical, building-oriented sustainability standards but the marketing for Sierra Morado focuses on the amenities without mention that Sierra Morado is part of Civano. In fact, a search on Pulte's website for "Civano" yields zero hits (as of 3/1/09).

**Laros:** What do you think about Pulte's "Sierra Morado" branding effort?

**Witmer:** It's amazing that they changed the name. I remember when that happened, when they came out and said they were going to change the name. Did they just buy the land for the land? They meet all the criteria of Civano, but decided to separate it.

For some reason, Pulte's business managers decided to separate Sierra Morado from Civano as much as possible. Pulte's advertisement media does not mention the Sustainable Energy Standard or the monitoring process which shows how the homes are performing much better than homes in Tucson at large. They certainly do not qualify their homes as green and a search for the word sustainable on their website yields zero results. Rather, they focus on items such as energy savings as one in a palate of amenities they offer as part of the Sierra Morado housing package. Pulte, at the time of this writing, was selling approximately 20 percent of the new residential construction in Tucson without green as a part of its marketing mantra.

When a consumer comes and talks to the Pulte sales team, they may get more in-depth information about the homes and even the Sustainable Energy Standard. Still, realtors may do their best to avoid expressing in any way that they have unnecessarily added costs to their homes; believing that consumers have the perception that green housing must come at the cost of the green they are trying to keep in their pockets. Really, the marketers of Sierra Morado may have simply decided that there were more people who would potentially buy a seemingly standard product with a lot of value built in than a product that was marketed as green.

Pulte did have some constraints built in with the planned area development (PAD) agreement; such as the challenge of providing 20 percent of the project with affordable housing units while still maintaining the environmental standards of Civano. The flexibility to change with the market, to make housing less expensive by trimming non essential items during poorer economic times, allows home builders to keep afloat. It may be that the request for affordable housing that also saves 50 percent heating and cooling energy and 60 percent potable water is a conflicting factor for some builders. Although Pulte did not go easily, they rose to the challenge and figured out a way to meet these standards in Civano.

There were other factors, architectural and design factors, to which Pulte succumbed. They had a quota to meet for homes without the standard front-of-home garage. New urbanism tends to lead designers toward trying to make the front of the home more human-oriented than car dominated to aid in community interaction. Part of the hope of this design strategy is that people will use their front porches and yards more, that they will be drawn as pedestrians to the roads more, and that the auto traffic in those areas will be less. To some, the connection between people and their cars goes much deeper. Moving rows of garages from the front of the homes into rows of garages in alleyways simply makes people come and go via alley way instead of driveway.

Pulte had many challenges that every builder in Civano has dealt with. One primary example is the as of yet uncompleted Civano Commercial Center. Although it would be wonderful to be able to sell Civano homes as part of a village community, complete with commercial services, at the time of this writing this was not yet possible. The idea was there and the groundwork done, but the buildings are not in the ground so no promises can be made. Although access to bus routes and other transportation benefits is a goal of Civano, much of that pressure is actually displaced onto the City of Tucson to provide, or not. So, other than dealing with Civano's interior road and trail designs, the developers of Civano could do little more than provide room for a couple bus stops.

Pulte is unique to Civano because it is the largest developer to be on the project and is on a very different organizational level than the smaller, local and regional builders who participated more at first. The large, vertically integrated Pulte Homes suffers somewhat from the effects of organizational size. Although the authors are not privy to the details of Pulte's business organizational structure, having worked closely with them, one realizes that many of the changes being made to Pulte's processes in Sierra Morado to adhere to Civano standards were unique to Sierra Morado. When new individuals came to the Civano project from within Pulte's organization, they often resorted to "business as usual" from other Pulte projects they had been on.

A specific challenge for Pulte's adaptation to Civano was the learning curve for the superintendants. They had technology presented to them that they had not seen on the previous five projects they were on. This type of challenge is overcome rather quickly because after they build a few units, they usually have adapted by then. Other issues arose with Pulte's integration of their purchasing and design departments. Examples would be the accidental ordering of otherwise code-compliant windows which would not pass muster for the SES requirements, or drawing sets which had the correct R-values specified on the summary sheet but not on the plan details. The inertia that comes with a large company has its disadvantages, but Pulte has done a very good job at balancing the redirection of its inertia toward meeting, and in some cases exceeding, the environmental goals of Civano. But they were also expert in negotiating out those items which they viewed too cumbersome to meet.

Pulte had experience building various features found in Civano's master plan, such as roundabouts in the streets. But other features, such as walkways in between homes in the high density areas, are complete departures from anything they had done previously. They used native desert landscape and not what would constitute a traditionally manicured yard (outside of Tucson). However, a very streamlined process, such as Pulte's production home building process, often requires that specialty items not common to their product line must be contracted out to companies with the desired expertise.

This speaks to some of the larger challenges with ramping up new technology uptake. Using Sierra Morado as an example, this was the first time Pulte Homes had a solar water heater standard on every home in a community. They needed help to provide that service because, although they have an in-house plumbing company, they did not have expertise in solar water heater system installation. So they hired the Solar Store—a thriving Tucson business owned by Katherine Kent, the same engineer who did some of the studies that helped in the creation of the Sustainable Energy Standard.

Although Katherine's company rose to the challenge, Pulte's decision to satisfy the solar requirement of the SES by using solar water heaters increased the local demand by up to approximately 20 units per month—the number of homes Pulte was closing during a busy housing market. In general, finding qualified installers for the new technologies is a challenge—and one that is compounded by a spike in demand. If the increase in demand is generally a continuing trend, the training of new technicians and build-up of crews is only a happy consequence of growth in that given sector. However, if the spike is a one-time event, balancing human resource management with meeting schedules can be more challenging.

As such, large, vertically integrated companies such as Pulte, with mostly in-house employees, have to make logical decisions about when to move new technologies in-house. They will do this if the contract labor force cannot keep up with their schedule or if the new process they are using is to become a standard fare. Therefore, some production builders generally offer upgrades and options based on the ones that are most frequently requested. If an upgrade item is purchased so rarely that it does not make economic sense to support the company infrastructure required to provide that service, the upgrade item will no longer be offered. It stands to reason that it would take a very regular and consistent consumer demand for a company like Pulte to reach out and provide something like a solar water heater as a standard upgrade item outside of Sierra Morado.

In the end, the whole point is closing as many houses as possible, as quickly as possible, with as many happy customers as possible and at a profit. At times when people can purchase homes cheaper than builders can develop them, builders generally don't build. This is much the case during the time of this writing—the economy is hurting and it is an extreme buyers' market. However, when the supply of finished lots gets used up in a region and the market comes around, most developers will probably start testing the waters by developing smaller projects. It may be quite some time before they move on to large, master planned, high-performance communities again.

Master planned communities can be developed in smaller phases to negate some of the potential risk of a slow market or a housing product that does not sell well for other reasons. However, the utilities need to go in and the feeder systems and roads must be designed and sized for the full-scale scope of the development. So there are still initial design and construction costs associated with master planned development that cannot be avoided even using incremental construction phases. In a sellers' market, it is assumed that homes will sell quickly, which will merit quick build-out and return of the investment in infrastructure. So there is not as much of a deterrent to building master planned communities during good housing growth times as when housing costs are down and selling new construction is a challenge. It could be devastating to a project's bottom line if the builder is forced to sit on undeveloped land with a heavy infrastructure investment already in the ground—one of Civano's lessons learned.

There were many community and home design features that Pulte eliminated from Sierra Morado (in comparison to Civano's neighborhood I) in an attempt to make its business model work, but there were some that they incorporated smoothly and across the project.

**Nichols:** Pulte constructed berms and swales to capture rainwater and planted low-water use plants. So even though they were not required to have reclaimed water, they didn't back away from the mandate to reduce use of potable water.

I expect they will actually do better than the phase I from an energy standpoint because they're a consistent builder. Phase 1 has been a hodge-podge of builders; some of them enthusiastically built better homes; others didn't. It's a mixed review. Phase I is interesting in that we had seven different builders, 50 different models. You know, it's a lot more interesting subdivision because of that. But Pulte usually builds thousands of homes per year, so if we can convert them, we're meeting the goal of the project. Because the goal of the project was never about the project itself, it was about showing—not demonstrating, but showing—profitability of building high-performance buildings and communities.

The emersion of Pulte as master developer, with its own set of core competencies and constraints, was a mixed blessing to Civano itself. But Pulte's involvement was a grand success in terms of the big picture of Civano—a master planned community intended to showcase an energy efficient, environmentally conscientious, and replicable development. One of the keys to a project being replicable in a commercial venture is profitability. So for Pulte to take cues from the Civano model and provide efficient homes with solar energy at little or no extra cost, it has to have been a success. Time will tell.

## AFFORDABLE NEW URBANISM

For mass production techniques to be adapted to include new urbanism while achieving the same economies of scale that have become ubiquitous with affordable housing, integration of the most valuable aspects of new urbanism must be part of the design vernacular and builder skill sets.

**Nichols:** Doesn't new urbanism mean being able to walk to the resources you need? So you would want to look to the commercial center and understand connections from there?

**Rayburn:** This is one of the big problems. One focus of new urbanism is to say, look, we have a transportation system, but it's car dominated, but it's there. So whenever possible let's think about using the establishment and building new urbanism around it. And if we can get mass transit in there, that is fantastic. If I were doing this, just talking about the east side here...let's pretend like there hasn't been an [economic] crash. Houghton Road is becoming a major thoroughfare. There will be a hospital there. There is already shopping. It makes absolutely no sense to try and build a shopping center in the middle of phases two and three [of Civano]. What does make sense is to work with the city to get a bus route, or tram, or something that connects us up.

I think you can create the look and feel of a walkable community where you then use public transportation to hook things together so you don't need to go into commercial neighborhood centers over and over again. What we were going to do in phases two and three was to decide that the neighborhood center doesn't have to be commercial. It could be a school, a library, a recreation center. It's not about having offices at the center, it is about having a place that takes a form that enhances your community experience.

There is little reason why a builder could not have a standard neighborhood center design, but part of the equation is to site the commercial center in a realistic location that links it to a larger commercial consumer base. This will require adaptation of how the neighborhood center connects to its surroundings. The Civano neighborhood I model put the intended commercial center at the center of the neighborhood—away from the main vehicular thoroughfare of Houghton Road. This choice drastically affected the way the roads and pathways move through the neighborhood. From all directions, the center is accessible in nearly equal capacity by walking or vehicle.

In a car dominated city, most retail and recreational businesses will need to be located within sight of the roads most often travelled. One of the first commercial businesses to open in Civano was a mainstream coffee shop. It seemed like a sure bet, located in a small neighborhood in a very classy development. It had a captive audience of Civano-ites hungry for local resources. But the sad truth is that this mainstream coffee shop was modeled after places almost always located in strip malls or grocery stores that cater to the impulsive motorist.

It is a game of averages. It takes a large number of people to go by before one comes in for a java. However, since the vast majority of people in Tucson go by in cars, to get the averages required to keep the shop open, it must be located near where the cars are or it must be next to another draw which brings motorists off the road. There are no pedestrians along the main thoroughfare of Houghton and the shop could not survive on local Civano walk-in business alone. So, unfortunately, Civano's little coffee shop died shortly after opening. It is very likely that neighborhood I's commercial center will not be a successful retail stop without completely unanimous community support or very strong draw from outside Civano. However, the center can be home to professional offices, the types of destination businesses that do not need to be located where the crowd is, but where customers go as part of a non-impulsive plan.

A developer may create a production-oriented community center design, but it is more likely that they would have to carry several versions of their design to allow for positioning of the center along differently oriented transportation corridors in ways which will allow them to plug in to the existing infrastructure and the new housing which will surround the center. Essentially, every new urbanism design that interfaces with an established urban setting must be minimally customized to suit the situation. It is unlikely that the new urbanist agenda can be bottled the way building codes are, but it is very likely that the process can be streamlined somewhat (See Chapter 9, LEED—Neighborhood Design). Either way, the increasingly complex up-front development requirements for providing well-planned communities sits largely on the shoulders of developers—who generally make their money by selling lots.

Lesson learned:

**Rayburn:** If you're a land developer, make sure you get a piece of the building pie.

The builders stood in line to make money whereas the only thing the land developer could make money on was the sale of the lot. For a developer to do an even more realistic, profitable, and cost conscious neighborhood I, there is a [significant] increase in management time, which transfers to a cost. This assertion is based on having gone through this and having met others who have done new urbanist projects.

At least in the case of Civano, the developers took on a lion's share of the risks. They had to have the patient money and they had to be the true instrument of hard fought change. This is not to take away from the efforts of all the community members and City of Tucson employees who worked to preserve the Civano values or the builders who rose to the challenge of meeting the project requirements. However, it is clear that the type of integration that is required to develop a new urbanist community requires more effort up front of the home building and that the greatest portion of that investment is made by the developer.

## SPEED OF BUILDOUT

Civano was expected to build out much, much more quickly than it did. The slow build time has been a major player in determining what size developer could manage the project. For the developers, interest accrued every day that the lots sat unsold. The smaller initial development companies ended up leaving the project a bit bewildered at the difficulty of the challenge they had attempted and some of the types of resistance they had encountered unexpectedly. Even the immense Fannie Mae became entangled in aspects of ownership that seemed unintended.

The homes in Civano came well before the 45-acre commercial center by over a decade. Functionally, this was never intended to be so. Originally, it was thought that key aspects of the commercial center would develop before, or in tandem with the housing, with the rest to follow shortly behind. This would have helped create the village dynamic of having a significant number of onsite jobs available as people populated the homes. What ended up happening was the creation of what will essentially function as a bedroom community to the greater Tucson area until sometime after the commercial center is fully completed. This clouds the image of Civano and many people don't realize the full potential of the project because it moved so slowly.

Excerpt from: **Impact System Monitoring Report #13:**

Community of Civano Neighborhood 1 & Sierra Morado For the Year 2007 Gallagher Witmer, Architect LLC, Distributed: July 21, 2008.

**Commercial Pavilions—"Civano Town Center"**

Planning and design began in the past years in the area known as the Commercial Pavilions/Civano Town Center. Both the developers of the proposed Rincon

Community Hospital at Civano and the proposed 7.5-acre retail development area along Houghton Road continued to meet.

Community meetings continued throughout 2007 with the designers of Tucson Medical Center's Rincon Community Hospital at Civano to be located in the lower third of the "TMC South" area. TMC intends to build a 123-bed community hospital, medical office buildings, and other medically related facilities. The original schedule has been extended to 2012 due to rising construction costs in the past years; TMC does expect to begin paving, utility, and site lighting improvements in early 2008.

Discussion and planning also continued regarding the development of the 9-acre area along Houghton Road. The developer, Jump Ventures, held a mini-charette with members of the Civano Neighbors Leadership Team in January to develop guiding principles and meeting with the larger community. Other topics raised were view orientation toward the Rincons and a village feel to the project. Also, in early 2007 approximately 960 consumer preference surveys were mailed to all the residents of Civano and Sierra Morado. The survey was also published in the Town Crier, the Civano Neighborhood Association newsletter. The goal of the survey was for the developer to gather input for the types of retail businesses most preferred. A 22 percent return response was received and food-based retail, both restaurants and groceries, was the strong preference. Jump Ventures is expected to begin work on the utilities and infrastructure in early 2008 similarly to TMC.

The Commercial Pavilions area is expected to bring commercial services very close to the community. This development will need to be connected to the Civano 1 and Sierra Morado's biking and pedestrian pathways, responding both to the community and the vehicular traffic on Houghton Road.

**Laros:** Do you think the commercial center can tie these disparate neighborhoods together?

**Rayburn:** Don't know. It's too hard to know. In some ways I don't think so. If we had continued the major road pattern, maybe. But that is not there right now. When you go down to Sierra Morado it becomes very confusing. If that had happened, then yes, even the hospital or whatever else is going down there would have been a great connector for all the communities. Now I think it will be harder because there is no natural connection by car. What came from the neighborhood I charrette session was a sort of rigid grid system. Wayne was the guy who said, "This is not paying attention to topography and we have to soften it up a bit." And so the curves started coming in as Wayne's thing.

## URBAN VILLAGES

Urban villages are basically the idea of having community centers with surrounding houses, but not necessarily designed to the extent that a pure and ideal new urbanist neighborhood would be. Figure 8.2 is the conceptual drawing for the Civano commercial center. This is much like what was expected by many to come from Westcor's

**Figure 8.2** The Pavilions conceptual drawing. *Drawing courtesy of John Jump, Jump Ventures Incorporated.*

planning process for the HAMP development. Some people think that this approach would find the best cost and benefit balance between common urban development practice and new urbanism. The urban village may be much more like the old communities that have become the models for new urbanism than new urbanist communities are themselves.

**Witmer:** I live in a 100-year old neighborhood near downtown and I've been thinking a lot about how it is very sustainable. The streetcar is coming through it. We have shops, restaurants, a grocery, and a hardware store. I've been thinking about taking the LEED for Neighborhood Development template and laying it over my neighborhood to see if it meets the standard, and I think it is going to, when you start to look at it. Everybody lives in a 100-year old home that nobody wants to tear down because it's a national register. It's this whole question of how existing neighborhoods are going to become more sustainable.

I think Civano could certainly do that. Maybe it is not up to Pulte to do, maybe it's up to the people to form some sort of cooperative. And maybe once the commercial center on Houghton is built, let's see what that does. If it's built just to face Houghton and the backside is faced to the neighborhood, I mean, way back when I started there we used to talk about that and how it should be a two-sided development. The stores should face both ways to the neighborhood and the Houghton corridor. Maybe that will make the difference. There is still the issue of the jobs. That location, when I look at LEED-ND, that's what I see. There are so many prerequisites, so many issues with infill and somehow already being where there is a lot going on.

**Laros:** The Congress for the New Urbanism talks about infill projects being blended with the infrastructure that is already around them, to mesh with the urban landscape; and projects which are located outside the developed urban center should have a more defined boundary and contain services within, like a village or a town. It seems like Civano tries to lean toward the latter of these, even though it is in a growth corridor.

**Witmer:** Who's to say which way its going to go? What I've seen, in seven or eight years, is that there is so much growth out there that it could become the center of town, or another node. Just down the road, Vail, has changed and grown so much. Who knows what it will look like in 50 years?

Most cities have their old, sought-after neighborhoods that have become known as excellent places to live. This may be due to an original design that was timeless so people naturally appreciate the place. More likely, well built places 100 years old have changed somewhat with their surroundings. Businesses come and go, but some are successful and become part of the community fabric. Mass transit may come through and link the neighborhood to the surrounding city in new ways. Homes may be demolished to widen streets, or community action may help avert such an action. HOAs may form to help patrol weeds—in a friendly way, or not.

The important thing to remember is that even after the last nail is set and the lines are painted in the parking lot—the community continues its development. It is important to not passively question what a place may be like in 100 years, but to proactively make every attempt to provide excellent physical and social foundations on which it can grow to achieve a vision of what it should be.

## DID THE PUBLIC AND PRIVATE PARTNERSHIP HELP OR HURT?

Civano's public and private partnership was a cumbersome beast, but that does not mean that all public and private agreements must be. At the core of such an agreement is public oversight of the development process that is usually bound to public funding. This oversight may be no different in purpose than that of a capitalized partner like Fannie Mae looking over the shoulder of those handling their money. The project sponsor is not necessarily going to be the end user, but this role is served to protect interests. Public interests are often protected in a public and private partnership via an open public review process, whereas private interests are often decided by fewer individuals. Thus, public oversight processes often move slower than private ones.

**Rayburn:** The most expensive money that you will ever have in your life is that incredibly cheap public money. But there is no question that without the public funding, Case Enterprises would not have been able to do it.

**Laros:** But what about the stick—the enforcement? Didn't the public involvement help keep the whole thing on track at times when it could have otherwise derailed?

**Rayburn:** I would say, emphatically, no. But aside of what came from bank loans and private investors; the rest was public funding and in-kind money.

We had to have this money. This is where I tip my hat big-time to Kevin Kelly because he was able to go out and form a series of partnerships that gave us about nine to ten million dollars in outright money, help in kind, and money in kind. This project would not have happened without that. We needed it. But it came with a host of difficulties. Far better for us would have been if the project were a quarter the size.

A far better approach to a public/private agreement, which has been used successfully in other places like Albuquerque, would be for the city to clearly state its policy goals. In exchange for developers meeting those goals, the city can offer fast tracked entitlement processes and work with contractors on permitting. Call it a day.

An interesting concept is the idea that a public point of view could be developed and organized in advance of development in such a way that does not hinder or slow the actual project build time. Lee Rayburn suggests that an excellent form of public and private partnership would encourage the authority having jurisdiction to have a clear set of guidelines for high performance development, a sort of primer or plan book, and when developers adhere to this advanced performance criterion, they will receive fast-tracked entitlement and contractors will have educational guidance in attaining permitting. These are the kinds of agreements now being struck in jurisdictions all across the USA.

The City of Austin, Texas released its first edition of "Green Building Guide: *A Sustainable Approach*" in 1992 and that document was one of the resources that influenced the development approaches to Civano. The City of Scottsdale launched Arizona's first green building program in 1998. Pima County followed suit and opened the doors to official green building in 2008. The City of Tucson already constructs its own buildings to LEED Silver plus 5 percent solar (essentially the SES) and, at the time of this writing, was finalizing plans for launching its own green building program in 2009. These programs all seek to better the quality of life for people in their regions, as well as diminish the impact development has on infrastructure and the environment.

If a city can provide services to efficient buildings which require less water, energy, and road infrastructure while producing a smaller waste stream, carbon footprint, and ecosystem disturbance—why would it promote anything less? Although Austin's first green building program was a model for the planning of Civano, it offered few incentives to professionals enrolled in the green builder program. They did support enrolled members in finding products, designs, and consultants to build according to the program guidelines and advertise for green builder program enrollees. Now, most new programs offer some kind of process benefit to builders in the form of fast-tracked or discounted building permits. In the case of Pima County, they became a provider for HERS rating so their in-house inspectors could provide key verification for LEED and other green building projects and create further value in their program.

Another example of system-level integration across all aspects of a project is the involvement of all jurisdictional stakeholders. There was a resolution passed by Tucson's city council that would have granted process expedition for builders developing Civano as one incentive for participating in forwarding the Sustainable Energy Standard and new urbanism. Unfortunately, the plan's examiners and code officials were not represented enough during this decision making process and much to the builders' chagrin, when it came time to expedite, there was resistance in the city and from other builders. The idea was there, but the support was only partially developed. In fact, when Civano builders came requesting their expedited permitting process, the plans examiners did not necessarily even know that was offered. This was another lesson learned at Civano.

Now green building programs are moving progressively closer to providing very efficient public oversight to green building without encumbering, and in many instances with enhancement, to the building process. The process that seeded Civano was an early attempt fraught with learning and discovery. Someday in the future private and public interests will likely need to work together again in such a belabored way to push the bar on sustainability into undiscovered territory. But for now, the era of the "green building program" has arrived to try and make the high performance building process as efficient as the resulting buildings and communities. This will ideally help with bringing greenhousing to consumers more affordably, in the long run.

## DOES SIERRA MORADO MEET PROJECT GOALS?

It is a fairly unanimous opinion that while Pulte Home's Sierra Morado at Civano does not fulfill all the urban design expectations, it does certainly achieve the goals of the Sustainable Energy Standard, water use expectations, and affordable housing. It is likely that those negotiating the land use agreement for Pulte and Fannie Mae understood that they were reducing the cost and risks of the project by diminishing urban design criterion. However, without a clearly set language (design book) to describe new urbanist expectations, it becomes difficult to specifically blame Pulte for falling short of expectations. Most would agree that although Pulte did not rise to the full extents of the Civano vision, it showed that resource efficient housing could be built affordably, including solar water heating. This is a major success for Civano in general.

### Success?

**Rayburn:** Always understand how different groups define success. They don't all define it the same way. For Fannie Mae, getting a big national builder in, Pulte, was success. So, Pulte came in and we went out. I can tell you that by that time the city energy manager was at a point where they were relieved to have Pulte in, the development agreement was renegotiated and some of the complicated language came out.

**Nichols:** They were very resistant to the solar energy component. But they did end up meeting all the energy components, they have the solar, they meet the water standards. From an engineering standpoint, when we talk about Pulte, we still think it is working. What part did they change? Where specifically is the phase II part of the project diverging in the touch and feel?

**Rayburn:** In order of importance of a new urbanist or neo-traditional community: First you think of the hierarchy of roads—big roads, secondary roads, alleys, service roads, bicycle paths, walking paths. Also, you need to have centers.

I think most people hold in their minds a map of their world. You always know where you are in phase I. You look over and see the neighborhood center tower and know immediately where you are in relation to the other things in the community. There is a way of helping develop the mental map we hold, our environmental perception. Why does it work? I don't know. I know it makes me feel better. That is non-existent in Pulte's phase II.

I don't think we did the best job we could have here in Civano creating the best bicycle and pedestrian environment we could have. We get an A for effort but it fell apart and got a little gnarly at the edges. But the desert country walkways—there is nothing like that down at Pulte's [phase II]. I'm not saying you have to do desert country walkways because they are very expensive, they chew up land. If you go back to the specific plan, there is a road map, there is a bicycle map, and a pedestrian map and they are all interrelated—about three different systems that are complementary but separate. These things did not happen in Sierra Morado. That is one of the reasons why I'm a huge advocate for new urbanism because it creates the kind of environment where people feel safer and are safer. In the end, I'd say that Pulte's work at Civano is really good, but typical sprawl development. The style of homes can be found in San Diego, New Mexico, New England. But that is why Pulte is Pulte. They are very systemized.

One of the glorious things about Civano phase I, when you look at it, is that the houses look like they belong in Tucson. What is the value of that? Well, it is a nice thing that creates a sense of place. I came here to Civano to build a community. The energy stuff is something I had to do but I think it was a good thing to do. So "look and feel" was where it led from, but all the energy and resource conservation had to be there. But Pulte seemed to feel you could only have one or the other. I think they could have done it. I can't tell you if it would have been as profitable as their standard stuff, but they could have done it. I hated what happened with the planning process, I thought it was a crime. I also understood that there are some benefits to having Pulte.

Success is measured in various ways. From a new urbanist perspective, even the most profitable urban-sprawl development is not a success, but a damaging way to create human habitat. To a strictly business minded developer, housing products that sell many units at a profit is the most basic description of success. What about to the homeowner? Affordability is highly important, but it is more of a factor determining what the consumer will buy than what the consumer wants to buy. Interestingly, during a buyers' market, we get more of an idea about what people want to purchase, versus what they purchase when more financially limited. Figure 8.3 shows some of Pulte Home's affordable, SES-compliant housing practices. Al and Lee discussed the effects of "boom and bust" on housing choice from the perspective of buyers and builders alike.

**Figure 8.3** Row of homes in Sierra Morado, Civano phase II.
*Drawing courtesy of Justin Cupp.*

**Nichols:** Have things changed since the 90s?

**Rayburn:** One of the great benefits of the housing bubble is the variety of housing that suddenly popped up in Tucson. Before things really took off, it was your standard house—period. After things really took off you have architects doing metal stuff, great homes down in the barrio. People were building homes in places where you thought people would never buy a home. It was artificial, but it was interesting.

I think at this point, if we were talking about starting the "new" Civano, clearly everybody gets the energy part. It's a totally different world at this point. I think it would be tons easier to do at this point.

**Nichols:** Do you think one could get financing for a project this size at this point?

**Rayburn:** Well, that would be tough, not for a while. If I were doing it, Civano II would not be as big as Civano I was. I would not make it a 2500 home development. I would make it much smaller. I would align myself with really good regional builders. I would make sure that I had city buy in on the type of planning we were going to do. I would keep them out of your business otherwise.

**Nichols:** Will there be innovation? If the politicians back up the idea of innovation—what then?

**Rayburn:** I think there will be real resistance to innovation until we're out of this housing mess. Everybody is going to be focused on that dollar amount. I also think that new urbanism is tried and true. There is lots of documentation that suggests that if you build a new urbanist community, your sales are going to be noticeably better, if not emphatically better than your competitions.

## Social success?

The Sierra Morado development in Civano has a different HOA than the first Civano neighborhood. They are separately governed associations with different CC&Rs and methods of operation within what was originally envisioned as one cohesive community. So the social fabric of the community is fractured along the same lines as the urban designs—having been forged by some of the same forces. But there have been attempts from within the communities themselves, with the help of Pulte and long-time Civano residents, to make inroads via a Civano-wide neighborhood association.

**Nichols:** We are totally making this integration thing happen. From where we were a year ago, we have made great progress. A year ago, we didn't have anybody from Sierra Morado attending any association meetings, and then we had the first meeting ever in the Sierra Morado community center and there were only two attendees from Sierra Morado. That may have had more to do with coordination of the first notice going out, but everyone in Sierra Morado is getting the Town Crier. So, they're now aware of it. It may be about a 50-50 mix now, or pretty close. But I'd love to see a Sierra Morado resident on the neighborhood association board, in some capacity, and when that happens—it should spread.

Part of the planning for a master planned place like Civano has to be deciding how the HOA and CC&Rs will be set up to work with the idea of the place. It seems like too many CC&Rs are just "cut and paste" from one development to another.

The most active neighborhood members are most likely the ones who will end up on a neighborhood or homeowners association board of directors. These community leaders then really have some clout to help set the tone of the community functioning and even debate. For example, they may either be a group inclined to help politely inform neighbors what consists of CC&R weed violations as they arise and educate neighbors in a friendly way why a particular plant is considered a pest in the neighborhood environment; or they may decide to simply leave stark citations on door handles with nothing more than a weed notice and dire consequences for failure to comply. Each approach is aimed at keeping pest weeds out of neighbors' yards, but every interaction between neighbors is part of the fabric of community spirit. So the methodology of the neighborhood leaders really can have a ripple effect throughout the community.

# 9

# CIVANO'S DNA

## Leading the Evolution

*Life is a progress, and not a station.*

—RALPH WALDO EMERSON

In Chapter 3, the idea that "some design ideas are before their time" was raised and accumulated some evidence over several chapters; but the idea was not laid to rest yet. Perhaps no idea is too forward thinking. Perhaps no idea is ahead of its time. Saying an idea is ahead of its time is almost like saying "we can't" or "it's impossible." But these statements are pinned to a time and place, when and where "we couldn't" or "it was impossible." What if the very idea of "can't" is a myth perpetuated by adolescent future planning? As our society's planning abilities mature, perhaps "can't" will disappear and be replaced by, "it is only a matter of when." Figure 9.1 shows the early off-grid concept of the solar village.

While we attempt to learn the full scale and scope of our environment, our perspectives are shifting. We may begin to think that ideas are only "too far forward" if we don't see the steps in between where we are, where we are going, and what we have learned from history.

## LESSONS LEARNED

The Hohokam people, progenitors to Civano, have been settled in and around the Tucson region for approximately 1800 years. As their civilization advanced through many periods of development they transitioned away from a simple agrarian lifestyle.

As trade with (what is now) Mexico brought a larger crop variety, the Hohokam experienced a growth period in their colonies. They adopted new cultural attributes such as Mexican-inspired ball courts and artistic flourishes on their pottery. During the Sedentary Period, which started around 975 AD, the Hohokam people were still experiencing significant population increases. Their houses became more substantial, they planted more land, and they developed village-like communities. Not only did the Sedentary Period Hohokam expand their home building technology, they are recognized as the first culture to master acid etching technology in the region.

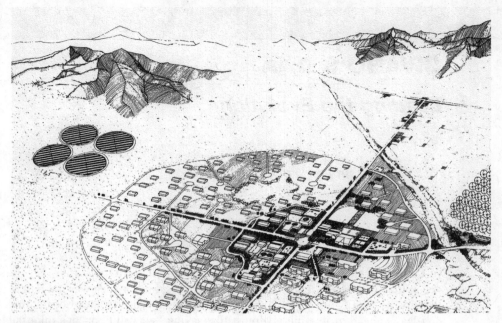

**Figure 9.1** Early concept: Civano as an off-grid village.

Then, between 1150 and 1300 (termed the Soho Phase), was the Hohokam civilization's pinnacle in terms of their overall population and urbanization. Their communities became more centralized and fortified. The many little farms and ranches became more consolidated, as did the canal systems that fed the fields. Political influence of the time between Hohokam villages became more predicated on the centralized management and control of water and other necessities. Those villages with more control over water access wielded greater authority. Hohokam architecture moved onto a much grander scale with resource-hungry four-story stone and adobe structures found in some villages. Unfortunately one never knows what nature, or one's neighbors, might do.

In the years following the Soho Phase, natural climate variances changed the carrying capacity of the people's cropland. Concurrently, the large centralized communities that had been developed over the past century and a half began to dissipate. A period of river flooding followed by a longer period of drought challenged the Hohokam's water management and agricultural technologies. At first, authority tightened and mobilized the people in an attempt to reengineer their canals, but nature did not comply. More flooding challenged the Hohokam beyond their ability to maintain their organizational structure—and centralized authority melted. This time period in Hohokam history, the Civano Phase, was a challenging time indeed. Sometimes it takes a barrage of challenges to make accomplished people do even better.

The Civano Phase marked a decline of the sheer extent of Hohokam development, but it was not a decline of their civilization. Faced with new social and environmental circumstances, the Hohokam adapted. Some stories say that a chief, possibly named Civano, helped foster a new way of thinking—a surge of innovation and a departure from earlier Hohokam traditions. They transitioned away from large buildings, they moved closer to the source of their water, they combined efforts with other people, they created more tightly formatted villages—they developed a new urbanism which harkened to earlier times, but with a modern understanding about why. The Civano Phase of Hohokam history is valuable to remember during the challenges of our modern times.

Sustainability? Sustainability is a phantom. Here and now sustainability is out of our reach. But that is exactly why we stretch ourselves and move out beyond our last hold to see what is on the other side—because we build our human environment just as our ancestors for millennia before have done. And we too need to reconsider how to persist and survive. "Too forward" thinking is merely far future planning and should play a vital role in how we make decisions today.

Gina Burton-Hampton spent some time with Al Nichols discussing what past experiences and events led him to become an energy engineer, and his unique way of progressive thinking.

**Burton-Hampton:** When did you first become fascinated with energy and solar power?

**Nichols:** It was probably when I was in fifth or sixth grade, about 1960, when we lived down on Bear Canyon when Dad (Dave Nichols) was playing with solar collectors. He was playing with polyester resin, trying to build a solar hot water heater, actually. He was playing with metal chips and things with a lot of surface area, thinking it would make it more efficient and collect more heat. He made a few of them. The first one he made was just a sheet metal box, about 6 inches deep. He had it setting out on the sawhorses in the yard with a thermometer in it. When he noticed it was about 300( in there, he ran in the house, got a bowl, filled it with beans and came out, lifted up the glass and set it in there—started to cook the beans.

Then there was the water out there. The well water was the worst water in Tucson. So that got me thinking about solar water stills. I didn't actually make one until about 1982.

It may very well be that necessity is the mother of invention. But this is a retroactive mantra. Needs today, or yesterday, beget invention today or tomorrow. But we no longer live in a time when we can justify retroactive behavior. Our knowledge is always expanding and with it our responsibility to act ahead of necessity as much as possible, if even only to plan for the time when the necessity will arrive. Maybe that is why the Nichols Solar Still, which delivers 'solar spring water' to a sink faucet to replace bottled water, is not a household item—yet (Figure 9.2).

The Civano project was pushed forward through many champions: owners, developers, financiers, and builders with different perspectives about what Civano was about. Some of the most poignant lessons learned during this process have been about how these leaders' involvement changed the course of Civano. In almost every situation, the friction created within this project was because of misaligned messages, incentives, and unmatched visions of what entailed success. Ultimately, these

**Figure 9.2** The Nichols solar still.

differences in perspective were the result of alternative ways of balancing needs of now with needs for the future. The early owners of the Civano project were so invested in the vision of a future development that they ultimately succumbed to financing realities and had to satisfy those needs by selling the project. Those who followed were better capitalized, but decided to change the vision of Civano somewhat to satisfy their belief of what could be sold at the time, so they could avoid the fate of earlier project champions.

It is likely that in perfect conditions, each and every person or company who is involved with the development of our human habitat would choose the path to a 100 percent sustainable future, and these authors imply no less of the developers and builders of Civano. The ideas of scale, integrated design, and systems-level thinking lead to a spectrum of ideal situations for a given task. Turned on its head, a given task is best completed with a spectrum of tools properly scaled, integrated, and systematized to do that job.

Pulte may have been too large an organization to provide the flexibility shown by the early builders of the first Civano neighborhood, but their production housing innovations fulfilled the needs to build a large scale high performance housing development with some aspects of a new urbanist community. David Case, David Butterfield, Kevin Kelly, and the first neighborhood builders were proactive, early innovators engaged in changing the community design paradigm. Although they left behind what is arguably a masterpiece of modern community-scale architecture in

Civano's neighborhood I, they required the added financial strength of Fannie Mae to attempt such a large project. These groups worked tirelessly with each other, with the public, and with the City of Tucson to find their balance. In some situations, the actors grew to meet the project vision, and in others the vision simplified to fit the actors, and the end result is Civano.

There will be unforeseen risks to even the most pristine development plans—at the very least inclement weather. More often the risks will go far beyond weather and no matter what powers of prediction we assert, we will come up either long or short of the ideal balance between the means wielded and the ends which we seek. Although it would have been ideal to interview each builder who ever set foot on Civano, one stands apart as a local Tucson builder who has been learning and growing alongside Civano and striving to find the balance between the means and the end. An interview with Richard Barna, Pepper Viner Homes, was an enlightening look into the experiences of a company trying to change the status quo, in anticipation of the future.

**Barna:** I'm Richard Barna, I'm with Pepper Viner and we're currently on our second project in Civano. The first one started about five or six years ago. At that time we met Civano standards, but we didn't really just dive in. We did what we had to do. We used solar and we learned a lot. The time and the mentality of the building industry then seemed like people weren't in a mind set to get into it.

The current Civano project we're doing fell right in line with a time of major change in our company. We realize that the measure of quality is really about how well the house works. So Civano was the perfect place to build houses which are healthier, more durable, more energy efficient. And we wanted to be able to measure it. That is key.

We want to actually get something out of doing that, in a business sense. So we've been fanatical about every step and part of the systems and the house, and how we put the systems together. This way we can understand how every little tweak to the systems affects the overall function and value of the house. So it's a great learning experience, but at the same time it is great for marketing purposes too because people can see it and we don't have to say it. We have people like the Department of Energy coming and giving us awards for how well we're doing. The IRS gives us tax credits for how well we're doing. It makes it a lot easier to express the value of your final product.

**Laros:** When did you do the first project in Civano? What was it called?

**Barna:** It was just Civano. We didn't have a name to it—they were just Pepper Viner homes at Civano.

Northridge, the new project, has its own branding name—not to separate itself from Civano, but because it was actually separated by the previous owner of the land and by the original platting of Civano. Even though we are in Civano neighborhood I, we have different rules and regulations set by the City of Tucson that govern that project. We did things there differently, not because we chose to do so, but because that was what was required by the plat. The plat is very clear about how it had to be.

**Laros:** We see some builders are very cautious about avoiding the use of the words sustainable and green in their marketing. Pepper Viner has a different approach. What has been your experience with that?

**Barna:** We do market green extensively. Our approach is to provide the benefits without increasing the cost. Maybe that was easier for us because our original product is a more upscale home. So there are costs in the original budget that we were able to redistribute instead of adding new costs. Without changing the original budget we were able to do things like save 50 percent on heating and cooling energy. So then you really want to market it by telling people that this house is really way better for the dollar. It would be suicide not to. So we have a giant mural in our sales office showing all the things that are different in our houses. We have our Burma Shave type signs at the entrance to the Northridge houses. We make a big deal about it.

And then, tying it in further, we couldn't afford to continue to build in the previous way in our other projects. Even though they were already bid out, we made the adjustment we had to make to bring those up to par. Maybe they weren't to the same level as the Civano project because it was difficult, once already contracted with the people involved, but we were able to take them to the energy star level. So everything we build is at least a 100 percent energy star level home. That is a big step and there is nobody else really doing that in Tucson.

So the philosophy for us is that we had better tell people what we're doing. It's a business decision, plus it's the right thing to do. We feel better about what we're doing now than we ever have because we're actually pushing building forward into a new way of building where the science of it is the key to making a better house.

Pepper Viner Homes is a local business building homes in Tucson and the immediate Region. Phil Pepper and Bill Viner are both graduates of the University of Arizona, both live in town, and are both in the office everyday. They are very hands on and very involved in doing things for the community. That's why this push toward green building in Tucson is important. We think it is for the community. It's for us, but it is a win, win, win. For us, for our customers, and for the community, because we can't continue to keep doing what we've been doing in this industry.

**Laros:** There is this underlying lesson learned at Civano about finding the developer, the financiers, and the builder who are all aligned with a common vision of what should be done. Do you have any perspective you'd like to offer about how the different neighborhoods of Civano have evolved and been developed?

**Barna:** I like Civano. We tout Civano to our buyers, but we do that for things like the ability to have a walking community and have community amenities which help people get closer to their neighbors. It's there, but it isn't a self-contained community even though that was part of the original vision. Almost everybody in Civano leaves and goes to work. So in that regard it would have been a lot better if Civano was close to the University of Arizona or downtown, but it doesn't matter. It still has a leg up over other communities. It is still valuable to our customers. It has a great school, walking areas, and a lot of people who are interested in doing the right thing. There was nothing wrong with the original vision—it was probably just a little out of place or a little out of time when it happened. If it was to happen again and start right now, it would probably be much more successful.

Pepper Viner Homes is a builder with its roots in Tucson and whose unique business vision is in line with a large amount of the vision of what Civano was intended to be. They are invested in the greater Tucson community, and that is where their reputation lives, and is on the line by their actions. But Pepper Viner, as any company, is on a learning curve. Perhaps the vision of Civano was not as clear in the beginning, but now parts of it are ingrained in the way Pepper Viner Homes does business. Does this mean that Pepper Viner could have taken on the whole, or even the remainder of the project? No—at least not yet.

**Laros:** Where do you think the green building movement is headed for the future?

**Barna:** When we look forward for ourselves, we build communities. We don't build them on a large scale because we're not a big enough company to build whole large communities like Civano. But every community we build, we think of as a self-contained community. We try to make it good for people to live in it. I think that will change some over the next several years where community activities, walking, recreation, the ability to reuse rainwater, maybe wastewater, maybe even the possibility of larger scale solar could come into play. I think that it is only going to keep going in that direction over time.

Right now it is hard because there aren't many new communities starting. We have a backlog of eight or ten communities that we haven't broke ground on yet. They are approved and ready to go, but they are on hold because of the economy. Even when the times change, there is going to be awhile before things get running. But things are so different than they were five or ten years ago.

We say there isn't a single part of building that doesn't have to be reconsidered. We look at everything. We relook at the way we wire houses. We relook at the way we plumb them. Our test house is set up to utilize gray water. We look at the way we put scuppers on so they could lend themselves better to catching rainwater if somebody wants to do that. We look at alternative building materials. We have a project coming up where everything is going to be built completely out of SIPs panels—and we're a production builder. So, things are changing. For us it is changing very quickly because we're committed to the change. But they will be changing for everybody. There will be no choice but to change.

So, a smoother path to developing a new Civano or the next great master planned, high performance community, would be to involve only a builder, or many builders with ethics and business goals which support the community, support the necessary skills for completion, and share the vision of a successful future. But even the most well rounded company faces challenges, and especially when it is trying to transform its organization, and must tread cautiously to maintain a safe foot hold on the bottom line.

**Barna:** Lessons learned are that every single decision you make when designing a home is interconnected. It is the same on the scale of a community. You have to keep everything in context of the final outcome and making decisions wisely. You can't waste money on something that just looks enticing, because those individual parts are just parts of a larger picture. You have to put your money to use wisely. This is a business. If it was an art piece without any specific budgetary constraints you could do whatever you wanted. But in business you want to get the best value for the money spent. In this case, that means you want the home that works the best.

An example, sometimes a client is nervous about investing in solar up front. So you put in the conduit and prep for solar, but then use the money in other areas to gain efficiencies. Then, if they should decide in two or three years that they want to add solar, it's going to compound the benefits. Those houses will be built so well that it takes less investment in solar to make a large impact on home performance. And the small cost and time for upfront preparation for things like solar and gray water makes a large investment in the future when energy will be more expensive and water will be more scarce than it is now. They'll be prepared.

No matter what we're doing now, we figure that five years from now things will have changed again. We don't want to lock in and think we've got it done. We want to keep it open-minded. There may be new materials, new types of solar. There may be new ways of approaching everything we do. Technology has turned its eyes toward building.

**Laros:** Does Pepper Viner expect more of its suppliers as a result of this push toward higher performance building practices?

**Barna:** Yes, we're helping one of our suppliers by participating in a test. We're switching to a higher performance concrete and documenting everything it is taking for us to do that. We're documenting what we went through with our ready-mix company who had to produce it and had never heard of such a thing, to our concrete contractor who thought we were loosing our minds. But now we have a concrete that is better and made with 40 percent fly ash, so it is a greener recycled content product.

We're also studying our transition from framing to SIPs construction and understanding what it means, what we have to do to satisfy the municipality, what we have to do to our blueprints, our bids, and all the trades who interact with the SIPs. By sharing this information with vendors who share these products all over the world, they can figure out what the obstacles are for builders trying to make this transition so they can learn and help builders go through the transition. In turn, they offer us technical support when we need it. We are demanding of them that they look at things and reconsider their own products, and they get from us knowledge from our experience transitioning within the industry. We are their view into the transition to green building, their opportunity to understand how their products work in the real world for the builders using them.

Although Pepper Viner, and the other homes of Civano, are under an auditing process, Pepper Viner adopted the strategy of employing scientific measurement to understand the performance qualities of the homes they build. This, too, is part and parcel to the ideology of Civano's Sustainable Energy Standard—a scientific, performance-based building code.

**Laros:** How do you measure performance in the homes?

**Barna:** There are two things—energy use and water use. We measure through the HERS (Home Energy Rating System) rating or something similar to the HERS rating. There are various rating scales. But the HERS rating is very good because it tells you what you are doing compared to an energy code home. It is on a 100-point scale, 100 being an energy code home and 0 being a home that uses no outside energy. They take it up to 150 or 200, which would be an older existing home built under older energy codes or no code.

We have one house, our sales office model, which is a 36 on the HERS rating scale. Most of our basic houses are in the 50s, which is about a 45 percent overall energy savings, pretty good. Then we have our test house that at the last count was at a 19. So we like to measure the energy. We like to look at every product we put in to measure water. The biggest issue with water savings is personal habit, but we try to outfit the houses with water saving technology. Now we're going toward the gray water plumbing and if somebody chooses to use it, that's great.

Material savings is a big thing. We really have taken note of what goes into dumpsters and what gets wasted. We look at that as a kind of goldmine of opportunity to make a house better, and use the money that is being wasted rather than just throwing it into a dumpster and paying for dump fees. This is really what drove us to using SIPs on our test home because that system reduces waste tremendously. They come precut and there is no dimensional lumber in them at all. So your whole structure is really using less natural materials and has almost nothing going into the dumpster. The only place you ever put any dimensional lumber in is where the structural engineer tells you to. So on some designs you may have a post here and there inside the panel. That is very minimal. You might have three or four posts in a house.

Then you look at the costs of the SIPs, which are more expensive than typical framing, but the insulation is included. That saves us time there. We don't have to drill it to run electrical. We don't have to cut sheetrock to fall on studs so that reduces waste. Sheetrock costs on SIP homes will eventually be reduced once the contractors catch on. The thing is, when you energy rate an SIP structure, it is continuous insulation. Even though it may only be R-24, it is a continuous, essentially unbroken insulated envelope. In framing, even if you pay for R-19, or R-23, you aren't getting anything close to that. The continuous value may be more like R-13, if you did a great job. Or you may have R-5 or R-6 if you didn't do a great job. That is the kind of thing we're measuring. So even if we pay a little more for SIPs, we end up ahead. Also, each time you save energy, you can diminish the cost of something like air conditioning to pay for the strategy that got you there.

So, that is the game. And it's really a good game because it really works that way. If everything you do and spend money on makes you end up with a better house for the same cost to customers, you win the game. Eventually, that will take over the market place because nobody can compete against that.

The Pepper Viner strategy as described by Richard Barna fits into the scale of its current housing product's price range. They started off selling a product at a given price, and then within that price range they have been learning how to evolve their practices to provide better functioning homes. They aren't cutting corners, unless the "corner" can be eliminated in exchange for something else which makes the whole product better. But, unfortunately, this product will still only be available to the upscale consumer to which Pepper Viner caters, or will it?

**Laros:** How does green building fit into the needs of affordable housing?

**Barne:** I've had people tell me they could never afford a Pepper Viner house. But if you look at the life cycle cost of a home, these types of homes are the best buys on the market.

The biggest problem is that the people who need it the most, because of the way the market is structured, they don't have the ability to get into the house that will save them in the long run. That is why affordable housing projects are so appealing.

We're about a year into a project where we were picked because we are going to build the most affordable of housing, super affordable, and do it green. It doesn't matter how many toys you can buy, what level granite countertops you install, or what level appliances you buy. We recognize that quality is a measure that has to be built into the house and no matter what price range it is, it has to be clear that it is built at a quality level. That is where HERS rating comes in, or reduction of waste. We're thrilled to be involved on a low-income housing project.

Now we're going to be building these efficient homes for people who need it the most. And once they get in, it will be completely durable, it will have concrete floors and a strong, simple SIPs structure, and they will be able to benefit from it forever.

Civano, still not fully grown, has already spread seeds throughout the community of Tucson and perhaps much further. New ideas have been spawned from the old, and some of the old are still maturing through new times. The story of a builder like Richard Barna's about Pepper Viner Home's journey is a true testament to why. Why to build a Civano. Why dare to reach, and why to choose a project that can be built on a scale that the local community builders can play a large role in.

## ARCHITECTS, ENGINEERS, AND NEW URBANISTS

### Parable

An architect, engineer, and new urbanist were sitting at a bar having a drink in celebration of completing their latest award-winning project. The bartender asked why they had been chosen for the honor. The architect said, "Well, as for me, it was the fine aesthetic qualities of the structures we built, their feeling of closeness with nature, regional architectural styles and staying within our client's budget!" The engineer was eager to say her part, "We specified used, ultra low-flow plumbing fixtures, high EER and HSPF mechanical equipment, solar PV, and solar thermal energy to complete the picture—form usually follows function and function is what wins awards!" And then there was a pause. The urban planner was looking back into the depths of the bar, seemingly in a daze. When he realized the silence was because his fellow project members were waiting for him to respond, he said, "Uh, oh sorry, what was the question? Um, I was busy contemplating how the room could be rearranged to encourage more people to use the pool tables and improve flow between the bathrooms and bar."

It is sometimes humorous to observe the discussions within design teams. After one has lived through a few projects, the stereotypical personalities given to different specialists become somewhat real. This is not necessarily because architects don't understand how engineers approach a project or vice versa. But they each serve a particular focus to get the job done and with focus often comes the necessity for a narrower overall view. More and more there is a new breed of design professional—people who are both architects and engineers or people who are specialists in human geography or project management. As the industry calls for deeper integration of the disciplines, individuals are reacting by attaining multidisciplinary educations.

Even though the overview of planning and designs of projects may be increasingly benefitted from multidisciplinary approaches, the measurement of performance is a focused discipline that must deal in the minutia of thermodynamics. Stories from a City of Tucson building official and then Al Nichols tell about how the source building codes for Civano became locally adapted. But these stories point to a hard line drawn in a multifaceted project that has generally succeeded in meeting its vision for increased home envelope efficiencies.

**McDonald**: It seems that there's a delicate interplay between the developer, builder, and the costs associated with adhering to the advance standards.

**Singleton:** Well yeah, obviously there is a little bit more cost involved initially up front in adhering to the sustainable energy code versus the IECC. However, in the long run, that gives him a sales advantage in that he can promote that he's going to save the owner on their utility bills throughout the life of the building. And the owner then is the one who really benefits from it, even though he pays a little higher cost up front probably for the structure, to start with. But in the long run, he saves on his energy bills. And if I remember correctly, Al Nichols indicated that his estimate was that they would probably recoup the difference in about three years just on the savings of the energy.

## The Engineer, in His Language

**Al Nichols:** We changed the heating-degree days, because we know it's hotter here than colder here, from 1440 to 2100 based on the fact that—

**McDonald:** I'm sorry. 1400 heating days? What is the heating-degree day?

**Nichols:** That's every degree above 65, or below 65, on average added up throughout the year. In this case, every degree below 65 per day equals about 1441. Okay, degrees in Tucson over the year below 65 times days is 1440. But we had measuring stations in the foothills. And because we had measuring stations in the foothills, we used that instead, which is 2100 degree days. And we made the Model Energy Code programmers make a special case for us.

We also knew that Mount Lemmon was a building site in Pima County. We didn't have any good data on it, but it's the same elevation, the same conditions as Flagstaff, which is 7000 heating-degree days; so we adopted 7000 heating-degree days for all those buildings above 4000 feet which were actually being built up at around 9000 feet on Mount Lemmon.

So we had a code that had 2100 heating-degree days below 4000 feet, code that said 7000 heating-degree days above 4000 feet. And to test if we had a house that was built at below 4000 heating-degree days standard through the Model Energy Code, again Katherine tested it on CalPass for 7000 heating-degree days. And we found that by using 7000 heating-degree days in the math modeling we approximately got 50 percent reduction in heating and cooling.

Paired with the fact that there's only a 5 percent difference in orientation, we said that we're close enough for all practical purposes to say we will build a house below 4000 feet that uses the same standard as though we were building on Mount Lemmon. Okay. And our results of that, that is the actual simulations that we've done—not simulations but our audits—using that 7000 heating-degree day standard and using the free and easy software from the Model Energy Code, which is now the ResCheck, which is free software, you can, in

15 or 20 minutes, you know, check to see if your home complies, which was the other part of making a standard that didn't require hiring a professional to do the calculations. In other words, as a committee, from my standpoint, it was—it has to be easy, it has to be measurable, and it has to be something that the builder can do themselves, that they can do their own calcs, they can take the building, they can plug in the numbers: I have a wall of 'X' R value, a roof of 'X' R value, I have got this specific kind of glass, I got an air conditioner with 'X' SEER rating, and it will tell you whether you pass or fail.

So using ResCheck, we created what they call the sustainable standard. And that was through the code committee. After we had done our work in the commission, we went back to the code committee and said we want to make the sustainable standard. We actually wanted to make it the Civano standard. But Dick Palmer on the—on the committee, a fellow mechanical engineer, didn't want to use the word Civano. He wanted to make it more generic. And which is fine because we didn't—we really didn't want it to be just the Civano standard. We wanted it to be a standard.

So the name came about as a sustainable energy standard. It was a hard fight. And just like when we first passed the Model Energy Code, it came through committee by one vote. And when it came through to vote on the Sustainable Energy Standard, right up to the last minute, Dick Palmer was fighting it. And because there was a council member in the audience, he changed his vote at the very last time—very last minute. And we passed it again by one vote.

But the debate was so lively; it was hard to tell who won at the end of the day. And they've always been that way. It was controversial because so many people thought that these energy codes would make homes unaffordable.

So we did that. The code committee endorsed the Sustainable Energy Standard as amended through our recommendations. On the commission subcommittee; I had a lot of people that signed up. But it ultimately only came down to two or three of us who wrote much of it because one of us was a building official at the time, or just retired—he knew how to write code. And so we would tell him this is what it's got to say, and he wrote the actual language. So a code official was the author of the language to put it in a legal format.

**McDonald:** So how does the Sustainable Energy Standard get applied?

**Nichols:** Well, first if you go to ResCheck and download Pima County's particular requirements because we have our own code. If you select the Sustainable Energy Standard, it calculates pass/fail based on 7000 heating-degree days.

But more importantly than that, there is the report and then there's the inspector's checklist. So if you click the report and the inspector's checklist, it prints out about four pages. And the first page shows your envelope, the building itself passes.

**McDonald:** Like R-values?

**Nichols:** The overall R-value of the building passes or fails and by a percentage. But in the ResCheck report checklist is all the rest of it—the specific language, including mandatory use of solar. There are about eight different methods of compliance—the designer can use solar in all these different ways. The builder has to have blower-door tests. You have to have done your manual J calculations. They have to, have to, have to, have to. All these things are the musts including not using fossil fuel to heat a pool unless it's a therapy pool. It's everything we put in there that made the project make sense. And those "musts" are the expansion of the Model Energy Code—the baseline code.

**McDonald:** So, you could say that Civano shows that one can build to the SES and mass-produced homes can be energy efficient?

**Nichols:** First off, when we passed the Model Energy Code, builders found that it was hardly noticeable. They often found that with a $300 to $500 upgrade they'd pass the code. When we passed the SES code, it was the same story. The builders that we presented it to were saying that the homes were going to cost 30 percent more. Reality is 5 percent at most. And we're seeing, now that we have numbers coming in from Pulte's Sierra Morado, which has very little architectural flair to it, the average resale price is about $3 a square foot or about 2 percent above market. In neighborhood 1, from the very beginning, these houses were about 15 percent over market. These are numbers we got from some of the local realtors who track this kind of thing.

**McDonald:** So that's right around $30,000 more for neighborhood I.

**Nichols:** Right. But you can only attribute about 5 perent or less to the energy standards, and the rest for architecture. So the architecture looks great and it's lovely but expensive. But the standard was written for affordable homes. And it wasn't to overburden. I mean, some folks expected homes to be off grid; you know, a whole subdivision of off-grid homes. We're just barely getting to that point right now that we can actually build an off-grid home, or a grid-neutral home, and those are horribly expensive. And no builder would have ever built them, okay.

But we finally convinced builders, and it was one at a time, to come and build to these standards. And after they ran the numbers, you know, now it's about $800 to $1500 more to add all the stuff. Solar collectors were subsidized at first, and we had a bunch of cheap ones. And they didn't work well; we had a lot of mechanical failures. The failures almost caused the collapse of the solar component of this project, which every builder has challenged or tried to get around. At the time, they would rather do almost anything but put solar energy in.

**McDonald:** Well, it's a good thing the builders don't make the rules, right.

**Nichols:** We made the rules, and we stick to the rules. Every, every builder has wanted to get out of that. But it's the Tucson Solar Village. And if a builder doesn't bring solar that builder can't build here. They can go build somewhere else. That's just been the hard line.

But we didn't ask for much. In fact, up until this very last amendment to the energy code, we never said by how much beneficial use of solar had to be applied. Okay. We never said 1 percent, 20 percent, or 100 percent. We left it to the good graces of the designers to pick from a list. Solar water heating became the most popular right away. But they had a choice of PV, solar daylighting, solar ovens, solar stills. You know, solar anything. Just something solar.

In this last SES code revision, we said the use of solar has to be at least 5 percent, period; just 5 percent. But for a residence, we figured that a standard 4 by 10 solar thermal collector, which works out to the equivalent of a 2000-watt water heater. There were a number of studies, and they came up with 5500 kilowatt hours equivalent per bedroom to make up 5 percent of the energy use in a typical SES home.

So up to four bedrooms you can still use a single panel and meet the code. If you have a five-bedroom house, you're going to have to have two. Okay. So we said you had to have that equivalent, and you can go on the Internet and find the equivalent kilowatt hours for all the solar collectors made. And if you use one of them, you can plot.

For commercial, commercial has always been calculated. There's no easy way to prove commercial compliance, you have to have an engineer. And you have to have engineers anyway for commercial.

And then everybody wanted a piece of it. And that's fine. You guys go, you know? But we got my little piece. And my little piece is codified and was passed by merit council. And the MOU, it says, you know, if you don't meet these standards, you know, it only takes somebody to complain. And if the complaint to the builder, the developer isn't satisfactory, then you go to the city manager. And the city manager has the right to cut off the permits.

So we had our stick. And you know, everybody wanted carrots. I'm going carrots don't work, sticks work. Even if it's against the law, people still break the law. But it was so clever that the city put in this monitoring report, because therein lies the stick. We had a very specific, codified, very easily provable standard; you know, very objective, not subjective. No, you can't use a tree to prove compliance. You know, tree dies; and then you lose your permit. Can't live there anymore. No, it had to be built into the structure. All this stuff had to be in the building itself, standalone. Any orientation, it all had to be just that way—and codified. Now the city did require that the model builders have their plans reviewed and professionally stamped.

But we've been able to defend the standards by just the fact that it's code; it's paragraph, subparagraph, exception … it's very specifically defined. And supporting software from DLE, which makes it free for the builder to check his own plans.

Per the Civano Memorandum of Understanding (MOU) 1998, Civano adopted the 1998 Sustainable Energy Standard (SES) for design and construction of all buildings in Civano. The SES identifies beneficial use of solar energy and a maximum use for hot water, cooling, and heating energy as 50 percent of the local standard as paramount to attaining a high performance level of energy use. Water use was also restricted. Through the IMPACT process, the MOU is adaptable by design and the current revisions to the SES approved by mayor and council on October 1, 2005 identify beneficial use of solar energy as a minimum of 5 percent while keeping the 50 percent heating and cooling energy reduction standard. The solar requirement has been met in most Civano homes with solar hot water heaters.

## 2005 SUSTAINABLE ENERGY STANDARD (APPENDICES A AND B)

Excerpt from the 2005 Sustainable Energy Standard: "The calculated target annual energy consumption of the building shell and mechanical system and domestic hot water heating shall be less than the energy required by the present Tucson/Pima County Model Energy Code by 50 percent. (Sustainable Energy Standard, Chapter 1, Section 101.4.)"

The Model Energy Code thereafter became the IECC when international standards were adopted; in this report, the Model Energy Code is referred to as the IECC.[1] Cooling and heating energy use by homes built to the 1995 Model Energy Code is approximately 36-54 kBtu/sq ft/year source energy. The 2005 SES proposed that energy use for homes built to the SES be 50 percent of the MEC, and therefore between 18–27 kBtu/sf/yr depending on the square footage of the 1995 base home. Evaluation of energy use was then to be evaluated yearly during the initial build out, as determined through energy audits of actual use. Small houses have more wall area per square foot than large houses, thus small houses tend to use more energy per square foot for heating and cooling than large houses.

The 1998 SES also described a need for "beneficial use of solar energy" but provided no parameters. This was rectified by the 2005 standard improvements. Solar hot water was most commonly provided by builders, but others relied on less rigorous criteria to meet this requirement which prompted the upgrading of the standard. The 2005 SES (October 1, 2005; Attachment B to Ordinance 10178) specifies the use of solar energy as 550 kBtu/yr/bedroom for residences and is prescriptively met using typical solar thermal hot water systems that have a rating of 2000 kWh/year or more for up to four bedrooms. Other means include PV or other methods allowed by the standard. Commercial buildings are to demonstrate a 5 percent utilization of solar energy.

SES compliance is determined by using a free DOE simulation program, ResCheck, to model Tucson residence designs at 7000 heating-degree days—the same number of

---

[1] ANE, Inc. reports on Civano Energy use for 2001-2002, 2002-2003, and 2003-2004 provide a history of the development of the 1998/2005 SES and its basis.

heating-degree days that is used for normal compliance in home designs to be built on top of Mt. Lemmon. This standard was determined by Kent Engineering by modeling homes in a more sophisticated program, CALPass, to achieve the 50 percent reduction in heating and cooling energy, and then putting the resulting home design parameters into REScheck and running simulations at different heating degree days until the same result (50 percent reduction in heating and cooling) was achieved. The DOE then modified REScheck to include an SES option that simulates designs at 7000 heating-degree days, specifically for modeling the SES in the Tucson valley. During simulation, Kent Engineering also determined that once a building envelope is designed to the SES standard, building orientation and aspect ratio have much less effect on energy performance (5 percent). The end result is a free, easy-to-use program that any contractor or home designer can download and use to determine if their buildings will pass the SES standard, without the more complicated concerns about home orientation or aspect ratio. ResCheck also provides an inspector's checklist for the solar requirements with each compliance report.

Energy evaluation of homes built in different years and per different energy standards potentially allows evaluation of the effects of codes and standards on real energy use. These results are important to stakeholders of Civano and to the City of Tucson. Broadly, evaluation of the SES and its methods helps to evolve conceptions and methods in sustainability. It aids the evolution of adequate (complete and correct) evaluation methods. The latter goal is explicit in Civano's Memorandum of Understanding.

The goal of the Memorandum of Understanding is to confirm the strategies for sustainable development and to implement and monitor the Civano IMPACT System ... subsequent monitoring of performance ... will provide the basis for determining the success in meeting the IMPACT system standards as well as the basis for improving future conservation and sustainability strategies and standards (Civano IMPACT MOU 1998, Sections 1–3; bold added).

## CONCLUSIONS FOR THE 2007 ENERGY USE STUDY

The results show a significant improvement in energy conservation under the Sustainable Energy Standard (SES) as applied in Civano homes compared to other homes across the city. The SES design certification (ResCheck) and the observed post-occupancy home performance correlate very well; indicating that the 7000 heating degree simulation is an adequate tool for designing to the SES standards in the Tucson climate. Furthermore, this study has shown that the requirements that achieve these savings are both financially and mechanically feasible for both the homeowners and the builders.

The era of the "green" building program is upon us. These programs help to streamline the types of decision-making processes that go into more environmentally conscientious development practices. Perhaps one of the most energy and time consuming processes of Civano was to define the goals and create the strategies to attain the goals. Now, project owners have access to tools that help them do that without the need to start from scratch or reinvent the wheel. The only program within striking distance of being able to manage a community scale development is the U.S. Green Building Council's Leadership in Energy and Environmental Design Neighborhood Development program.

## LEED—Neighborhood Development, Green Construction and Technology

The LEED—Neighborhood Development program was still in pilot at the time of this writing, but many of the strategies for green construction techniques and technologies had been developed within the USGBC for years, used in established programs, and are applied in the Neighborhood Development program. Civano's development considered many aspects of more environmentally conscientious building practices, but a comparison to the LEED Neighborhood Development pilot program requirements and points reveals the strengths of such up-and-coming programs at casting a wider net over a project, capturing as many opportunities to increase its environmental performance as possible.

### CONSTRUCTION ACTIVITY

The LEED—Neighborhood Development program has a prerequisite which mandates measures are planned and carried out to minimize site disturbances due to construction activity. Primary concern is given to soil erosion control, minimizing waterway sedimentation, and airborne dust generation. Usually the first construction activities, after surveying has been done, require blading and grading of a site. No matter how extensively, this must be done to clear for paths, roads, utilities, and building pads. This process is done with heavy machinery and from the time the first machine rolls onto the project site, plant and soil disturbance begins. As the buckets and blades hit the ground, site hydrology is altered. These things are essentially unavoidable. But the effects can be managed.

LEED requires an erosion and sedimentation control plan be completed which outlines best practices in preventing erosion and loss of topsoil during storm or wind events. Even if the crew stockpiles topsoil for reuse later, keeping the topsoil on site helps prevent sedimentation of downstream water systems and the stormwater conveyance systems in between. The erosion and sedimentation control plan also needs to outline how pollution of the air with dust and particulate matter will be prevented. Many jurisdictions already require compliance with some of these measures, especially airborne dust and particulate control. LEED requires that the best management practices for erosion and sedimentation control are selected from the 2003 EPA Construction General Permit (CGP) OR local erosion and sedimentation control standards and codes, whichever is more stringent.

## GREEN CONSTRUCTION

The LEED—Neighborhood Development program awards points for neighborhood buildings that become LEED certified. Civano was all comprised of new construction. If it had been a LEED project, this would have meant that a minimum percentage of buildings would have had to adhere to the LEED for New Construction, LEED for Homes, or LEED for Schools to start acquiring these points. LEED construction guidelines require new construction reaches a minimum level of environmental and performance considerations to even become certified at the entry level. Considerations of location and linkages, as discussed in Chapter 1, would be applied to each structure. In a mixed use, high performance master planned community design such as Civano's, locating most or all of the structures in high density and in close proximity to services and transit centers is already part of the design thinking, making these goals much easier to attain for the individual homes and buildings within the project.

## ENERGY AND WATER EFFICIENCY

There are different ways a LEED—Neighborhood Development project can earn points for energy and water efficiency, and different ways to document these points. In summation, the pilot LEED program recognizes three levels of achievement for energy conservation: 10, 15, or 20 percent improvement compared to the baseline building performance rating per ASHRAE/ IESNA Standard 90.1-2004. This must be achieved for 90 percent of project buildings by a whole building project—using LEED approved performance rating methods. There are several paths to compliance and verification for the LEED process, all outlined in the guidelines.

Civano used its own baseline from which to measure performance, but would have exceeded even a 20 percent performance improvement target example in the LEED—Neighborhood Development program. However, LEED offers the methodology up front for designers and builders to consider, whereas all of these processes and metrics had to be designed for Civano. Having been systematized and figured out up front makes programs like LEED very strong organizational tools for high performance development projects.

Water conservation on LEED projects also predicates awarding points based on saving water, by percentage, over a baseline quantity. LEED—Neighborhood Development delineates two different categories: commercial and residential structures over three stories, as well as residential structures three stories or less. This allows for different tolerances for the demands of commercial versus residential structures. Compared to Civano's performance goals of reducing potable water use by 60 percent over the city standards, even the most progressive water efficiency goals of the LEED program are no match. This is most likely a result of the increased awareness about water resource scarcity in Tucson's arid Sonoran Desert climate. In 2009, the USGBC took strides to start regionalizing LEED points, weighting items such as water efficiency according to the special conditions found in the project location.

Water conservation efforts in Civano have been largely focused on outdoor landscape efficiency. LEED—Neighborhood Development offers one point for irrigation which uses only captured rainwater, recycled wastewater, gray water, or reclaimed water treated and conveyed by a public agency specifically for nonpotable uses. The point can also be acquired by installing landscaping that does not require permanent irrigation systems. To help landscape plants take hold, irrigation systems can be installed for up to one year to help with plant establishment, but then they must be removed.

Civano uses only low water use landscaping (with the exception of a play field, two smaller grass fields, and some back yards) and all common areas are irrigated with city-reclaimed water. Rainwater harvesting was employed in the Civano landscape design by trying to slow and trap runoff in berms and swales. Many residents have been installing rainwater cisterns that store the excess water for later use, after the ground has dried. Figure 9.3 shows a schematic for a very robust and inexpensive rainwater collection system. Civano would have excelled in the LEED context for water resource management, and proves that even more aggressive measures are viable and attainable.

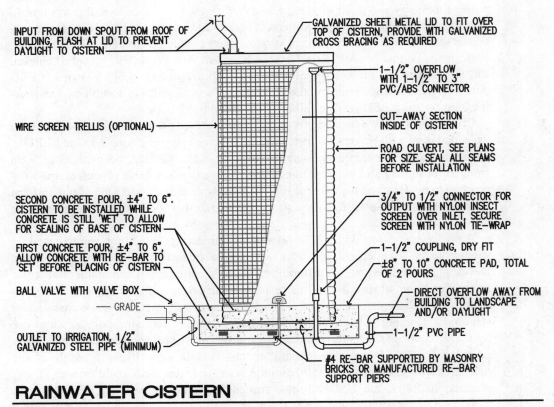

**Figure 9.3** A cistern schematic.

## MINIMIZE RESOURCE CONSUMPTION AND SITE DISTURBANCES

LEED—Neighborhood Development also promotes the reuse of existing buildings as well as preservation of historic buildings. These measures are very appropriate to diminish waste streams while conserving resources and neighborhood character on infill or redevelopment projects; but Civano was built as a new construction project and would have had no opportunity to acquire these points.

LEED also rewards projects that are built on previously developed land. The LEED—Neighborhood Development process offers two options to achieve this. One option is to locate the development so 100 percent of the zone of impact is on previously developed land. Civano had little to no previously developed land in its footprint. The second option prescribes how much of the development footprint must be undisturbed given the project's residential building density. Measures must be taken to ensure that the undisturbed areas will be protected into the future.

Civano used some innovative techniques to preserve the major flora on site by transplanting the large trees from areas being graded into the finished landscaping. There was also a mandate to preserve 30 percent open space in the project, which was achieved. These are exactly the types of measures the LEED—Neighborhood Development program wishes to promote, and for which it provides guidelines to do so.

## MANAGING STORMWATER

The LEED credo when it comes to stormwater management is to preserve or mimic the natural site hydrological qualities as much as possible. This includes minimizing the sediment and toxins going downstream from the project, as well as reducing runoff speed and potential for erosion of waterways. The pilot LEED—Neighborhood Development program suggests up to five points for this effort by designating how much water must be able to infiltrate, be reused, or evaporate given the amount of rainfall which will hit the project's impervious areas per year.

The LEED stormwater management guidelines are less strict for previously developed sites that may have existing impervious groundcover already in place than for newly developed sites. Although Civano was designed to direct a large portion of storm runoff into planted medians and other catchment and sediment basins, its design guidelines lacked this type of specific calculation to determine successes in stormwater management. However, as water resource awareness in Tucson grows, the community evolves and adapts new practices. In late 2008 the City adopted a new land use code which requires much more stringent water budgeting for all new commercial construction. By 2010 commercial properties will have to show 50 percent rainwater harvesting and residential buildings shall have gray water stub outs so homeowners will be able to utilize gray water in the future.

## SOLAR MANAGEMENT

When clients come to Al Nichols Engineering, Inc. with ideas about passive solar building design, they are met with open arms. But the message they hear is sometimes confusing at first because the meaning of "proper solar orientation" or "passive solar design" definitely depends on the context of the climate where construction is to occur. Most of the mainstream concepts about passive solar are about harnessing the sun, whereas in a place like Tucson the biggest concern is how to avoid it. There is no doubt that even in very cold climates, the sun can be harnessed and used effectively to heat a space—even to the extent where nothing more than a wood burning stove would be required for supplemental heat during rare extended cold events (see the "passive structures" section earlier in Chapter 7).

This is one area where a prescriptive green building program can have built-in pitfalls if the language does not allow for adaptation to multiple scenarios. It is somewhat difficult to create a prescriptive solar-heated (or heat-rejecting) home in a non-regional context because the language would represent a specific design strategy to figure out what approach is needed in each home configuration and each climate. A set of solar designed house plans could be developed for a given location or locations, or approved for a specific housing development. At Civano, there was ongoing debate about the approach to solar orientation in Civano. One model created rows of homes, all lined up in the same direction, on streets running essentially parallel. Although the grid system is prevalent throughout many American cities, this approach to solar design was deemed out of balance with some of the other goals of the community.

The other approach extreme is to custom tailor each lot and each home to fit in a more organic layout, with curving streets and a higher level of pedestrian comforts while still designed to be passively heated and ventilated. This exercise was out of balance with the affordability goals of the project at the time. However, as described by Al Nichols in Chapter 4, the studies leading up to the development of the Sustainable Energy Standard revealed that a very well insulated envelope and a building design with a shape close to square will be only slightly affected by solar orientation. This compromise helped bolster the argument for additional home envelope performance, while meeting some of the wider reaching goals of the community.

Although these compromises are in support of the efficiencies of production housing, they do not mean that passive solar cannot be delivered in an efficient production fashion. Passive solar design on a large scale is an excellent goal that adds some challenges to companies attempting to reach mass production economies of scale. One solar design aspect of Civano that has succeeded nicely in the first neighborhood is the attempt to reduce the heat island effect. Heat island effects are the accumulation of thermal energy in person-made landscapes that create a local rise in ambient temperature above adjacent undeveloped land ambient temperature.

LEED strategies are designed to identify several different approaches to minimizing the heat absorption qualities of the development footprint by maximizing the aggregate reflectivity (albedo) of the landscaping and roofs of buildings. The feature of our built environment most often accused for artificial heating is blacktop street paving. The blacktop is a dark, massive substance on top of compacted rock and earth that is a very effective thermal heat sink and energy storage device. There are even some paving projects in the world which double as large solar water heaters, harkening to the days when Romans used black slate stone to accumulate solar heat and transfer it to water passing over them into the baths. The heat island effect is very pronounced in cities such as Tucson and Phoenix, but it is a concern even in climates less heated than southern Arizona's.

Civano's first phase had narrower streets lined with native trees, saved from the building site and replanted between nearly every sidewalk and streetscape. The combination of reduced blacktop area and consistent shade helps keep the walk in the mid summer temperatures much more bearable than a ramble down Tucson's Speedway Boulevard. The City of Tucson required that all alleyways, where garbage and recycling are picked up, were paved. This added to the originally planned amount of neighborhood heat sink, but there is strong argument that the reduction in dust, mud, potholes, and weed control is the type of trade off which makes pavement attractive.

There are alternative paving strategies and materials with higher albedos (measurement of reflectivity), but none yet as inexpensive and ubiquitous as asphalt. However, with ingenuity future developments trying to solve this same dilemma may find a different compromise such as concrete or perhaps even well-bedded, lighter colored cobblestone. Cobble would allow more on-sight infiltration of rainwater. The labor costs for such a feature would have likely been very prohibitive, but other advantages could be gained from using an alternative material. Cobble is also a material that does not require very large machinery to pull up in small sections and re-set, and the stone itself can usually be reused.

The parts of Civano that do not closely follow the first neighborhood's urban design and landscaping cues are less protected from the sun. The roads are a bit wider, there is less plant life in the medians, and the trees are spaced further apart. The later development areas of Civano did not get the benefit of the large stock of harvested mature trees, but they did plant many young trees that will become effective living shade devices in several years of growth.

The builders of the later phases of Civano reduced the amount of planting in their medians and sidewalk buffers because the first neighborhood had very high, and disputed, maintenance costs associated with the lush landscaping. Although the later Civano neighborhoods will be notably warmer places to walk than the first, they will still be a far more comfortable place to walk than the comparison stroll, Speedway (heat island) Boulevard. A place like Civano could encourage redevelopment of places like Speedway Boulevard to be retrofitted with curb cuts designed to harvest and filter water into planted medians instead of shuttling it away, dirty and downstream.

## Tree shruggers

The use of foliage to provide a more suitable microclimate is a difficult idea to knock down at all, because the idea of planting trees is symbolic to the environmental movement. Plants remove carbon from the atmosphere, add oxygen, and provide all means of other aesthetic and functional benefits to society. Regardless, even regionally adapted plants are benefitted from maintenance and there are people who don't remember to water them during a drought. There have been droughts so bad that in some parts of the world people had to choose to drink over saving a tree.

Furthermore, there is a cost to owning a landscape like Civano's, watered with city reclaimed water. Although the reuse of a resource like water is a key strategy to conservation, nearly everything requires an input of energy and resources to recycle or reuse. The cost of the infrastructure to deliver reclaimed water is immense, and the pumping of the water creates a large electricity demand.

These are not reasons to stop planting, but they are part of balancing the equation so that our natural resources are impacted as little as possible in aggregate. Landscape flora should be considered a good strategy to enhance the energy design of a home, but it should not be a way of meeting an energy or heat island code in and of itself. There is no way to regulate what a person will do in their own yard and a person who appreciates a tree for its interaction with the microclimate of his property may sell his home to a person who would rather have an open view. The latter has no problem saying, "Goodbye tree, time to re-enter the carbon cycle in the form of smoke from my chimney." But what of the energy code which relied on that tree to be effective? It is up in smoke as well.

Just as infrastructure-demanding reclaimed water has proven more feasible for watering larger sections of shared landscape than individual yards, landscape maintained by a solid homeowners association, neighborhood group, or municipality may be a more valid arena to employ floral-based heat control strategies. These entities would be more likely to consistently maintain the landscapes, keep them watered during droughts, and maintain a consistent point of view about the value of those landscapes to the community.

# Onsite Energy and Renewable Energy Generation

LEED—Neighborhood Development pilot rating system offers one point for onsite renewable energy generation that meets that of the Sustainable Energy Standard. The performance-based target is to produce 5 percent of the overall community energy consumption (electrical and thermal) by use of onsite renewable energy resources such as solar, wind, small scale hydroelectric, geothermal or biomass. Another point can be gained by the capacity to generate 5 percent of the community's designed peak electrical load onsite.

Civano's Sustainable Energy Standard allows credit for onsite co-generation in place of the 5 percent solar energy requirement. Co-generation is the creation of electricity from a fuel source, such as natural gas being burned in a turbine generator, and using the waste thermal heat to warm water for various processes. This is a very efficient, dual use of the energies required to turn fuel into electricity. In Civano, however, the co-generation processes must provide 5 percent increments of energy requirements to the subject facility for every 1 percent increment of solar which may be removed from the requirement. There have been no co-generation facilities built in Civano at the time of this writing, but the promise of a future hospital gives a solid opportunity for such an event on the horizon.

Co-generation is often used in very large facilities, or complexes, where three general conditions exist.

**1** There is a large, consistent electrical requirement.
**2** There is a large, consistent thermal heating requirement.
**3** There is a readily available fuel source.

This is why a hospital may be a valid choice for co-generation. The initial investment in such a system is much higher than simply plugging into the existing grid resources because of the complexity and size of the infrastructure and equipment needed. But the increase in efficiency is significant. The approximate energy cost to deliver natural gas is much lower than electricity. But the benefit of paying more for grid electricity is because all of the generation equipment exists as some other entity's responsibility. Locating the electrical generation equipment at the point of use can allow for a more efficient consumption cycle. Furthermore, the waste heat from the production of electricity at the power plant cannot be used to deliver thermal energy at too distant a point of delivery. Therefore, in the right conditions, the efficiencies of onsite generation can be compounded where waste heat can be utilized, thus recovering some of the energy from the process.

## DISTRICT CENTRAL PLANTS

A central plant that provides hot and cold water to a facility works on the same principals of scale as a co-generation facility. A hospital could have many smaller water heaters on each floor with all of the associated plumbing required to connect them to the fixtures they serve. If they are efficient, closed combustion gas heaters they each require wiring, a gas line, combustion air intake, and an exhaust flue. If the water heaters are electric, they require wiring but will generally use more money to provide the same Btus as gas heaters.

Similarly, air conditioning and heating for the spaces would be provided by either a large packaged, zoned HVAC system, or several smaller systems in different zones around the facility. One major benefit to these formats is that each water heater or HVAC component, taken individually, is fairly inexpensive to replace and service compared to a central plant component. But each of these pieces of equipment is also a point of efficiency loss and material consumption redundancies.

When there is a consistent need, all of the cool for the air conditioning, heat for the water, and warmth for the space can be provided by one centralized facility. The basic primary components of a central plant are essentially a large efficient boiler, a large compressor, and probably a cooling tower (to help with heat rejection). The redundancy for a central plant is usually singular, meaning there will be one back up of each major appliance capable of at least providing the baseline requirement of the facility it serves. This allows for the omission of using many less-efficient water heaters as well as various components of the HVAC system. These would be replaced with a campus piping system, a couple large tanks, pumps, many valves, and heat exchangers.

This description is lacking much of the complexity central plants actually entail. Some state of the art facilities are highly refined and can produce accurate, process-grade thermal gradients for industrial applications. While small stand-alone unit efficiencies are on the rise, the overall efficiencies of a well conceived and operated central plant in providing heat and cool for various uses is still superior in general compared to stand-alone strategies. Civano has some central cooling in the first neighborhood community center, and the LEED—Neighborhood Development pilot program offers points for projects that utilize these strategies at least to a minimally prescribed amount.

## WASTEWATER MANAGEMENT

The LEED—Neighborhood Development pilot program offers one point for projects that can demonstrate at least 50 percent of the wastewater that would otherwise have gone to the sewer is redirected to uses onsite to replace the use of potable water. Civano has no onsite wastewater treatment facility, but rather, nearly every building is connected directly to the sewer. Although Civano would have fallen very short of earning this point had it been a LEED project, there are examples of graywater use in the community. Al Nichols Engineering, Inc. is a mixed-use facility with office use downstairs and residential living space upstairs. The residence diverts a significant percentage of its wastewater stream to the landscaping surrounding the office entry, the courtyard, and alleyway landscaping.

Although the flow has not been measured specifically, the showers, one lavatory, and the clothes washer all are connected to the graywater system. This only leaves the kitchen sink, one lavatory, the toilets, and the dishwasher going solely to the city sewer system. The ANE, Inc. gray water system is a gravity-fed system, which is one reason why only the upstairs residence serves as a grey water source. The other reason is that the office has a very low water use, and comprises a small percentage of the building's overall potable water use.

Gray water systems require attention and maintenance. They can be as simple as a line that directs water from the clothes washer or shower directly to the landscape it is watering, or as complex as to require storage tanks, filtration, a pump, pressure tank, irrigation timer, and plumbing. However, even the simplest gray water system tends to carry materials, especially synthetic fibers, which eventually clog the soil and limit infiltration of the water. This requires the homeowner to aerate and turn the soil or completely relocate the system discharge to a new place. Homeowners also need to have the wherewithal to be careful. Gray water should never pond or pool. Rather, it should be delivered below the surface or only to ground that will soak up the water quickly, or it can turn septic.

Higher levels of bacteria and other substances are part of the challenge to managing gray water. These are not insurmountable challenges, and as homeowners become

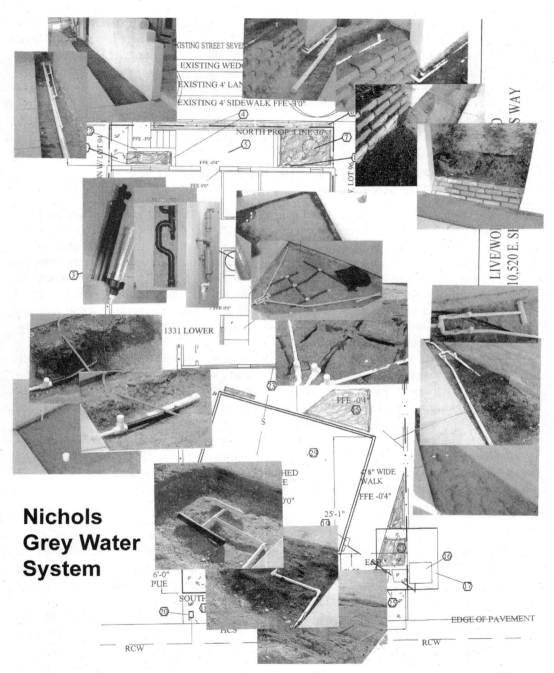

**Figure 9.4** ANE, Inc. gray water system.

more savvy, neighborhoods attempting to achieve potable water use reductions could design gray water systems into the home designs. However, places like Wayne Moody's Milagro Cohousing have a neighborhood scale wastewater treatment pond. This approach is a very local version of a city reclaimed water system. All the houses' waste lines go to the same place for processing and the reclaimed water, once satisfactorily treated, is delivered back for use in the landscape. This requires much less homeowner knowhow, and it can even create an amenity for the neighborhood by growing plants and attracting birds and other wildlife.

## RECYCLED CONTENT IN THE INFRASTRUCTURE AND WASTE MANAGEMENT

The LEED—Neighborhood Development pilot program will award a point for projects that diminish the percentage of new raw materials that are used in the roads, sidewalks, parking lots, and curbs. Civano was built without that specific mandate which awards strategies such as using 90 percent reclaimed and crushed concrete as sub-base aggregate in road construction, for example. There are other means to reduce the amount of virgin materials which go into the infrastructure material palate, such as using fly ash in concrete or scrap tires in asphalt paving. The gains of using recycled or remanufactured goods in a development strategy would be overwhelmed if the project produced too much waste.

The LEED—Neighborhood Design pilot program also can award one point for projects that divert a significant portion of the waste stream out of landfills and into recycling programs. A project can comply by providing various organized means for residents to easily recycle, such as being part of a city-wide recycling program or providing recycling, reuse, or composting pick up and drop off points available to anybody in the community. The City of Tucson serves Civano with its blue barrel recycling program. Every home and business has a green garbage bin for non-recyclables, and a blue bin for products that can be recycled. They are picked up on the same day each week.

## DARK SKIES

Tucson is in close proximity to several astronomical observatories in the region. There is a dark sky ordinance that imposes limits on the type and density of lighting that can leak from a project. Civano is a fairly dark place at night, lacking the common rows of towering streetlights which are common in some neighborhoods. Most of the fixtures are designed to cast light down onto pathways and areas most in need of light, rather than illuminating whole areas. The LEED guidelines recognize nighttime glare from lighting to be a kind of pollutant that reduces access to nighttime skies which can diminish the enjoyment of residents and impact nocturnal environments.

The Leadership in Energy and Environmental Design sets the stage for teams to streamline design based on known strategies and measurable baselines. The LEED programs themselves are somewhat like plan books. They offer design guidelines that give builders a realm of choices and helps them understand the relationships, consequences, and constraints of each design choice they consider with regards to the integration of design choices intended to improve the environmental performance of the resulting project.

The Civano process led to a project wherein the ultimate performance goals and achievements of the community would have likely garnered many of the LEED—Neighborhood Design program points at the maximum level. Civano also has very innovative aspects to its conception that would have likely increased the points it could have received as a LEED project under the "Innovation and Design Process" category. The ongoing adaptability of Civano due to the audit process is unique, and provides incredibly valuable insight into green building practice.

More and more, LEED project certification is being mandated for government buildings, universities, and even on U.S. Air Force bases as a device and methodology to ensure projects are more environmentally contentious; but also that they cost less to operate and own. They are typically healthier to live and work in, more durable, use less energy and water, and have value built in which cannot be found in less scrutinized efforts. Whereas truly groundbreaking projects for their time, such as Civano, need the full court press of a public and private collaboration, the LEED process and other green building programs are there for people to use as the practices which yield better buildings and places have become ever more refined.

Civano's final goal-oriented focus did not put as much scrutiny on the process of the construction itself as a LEED project might. The types of decisions which led to which site Civano was built on were driven by different constraints than a strictly private project would have been. The production builders of that time were among the first in the region to take on the high-performance building learning curve. Although Civano has taken shape, and the foundation has been poured, the journey is not over. Civano's DNA is that of a high-performance community coded with knowledge handed down since before the Hohokam Civano Era, and it can continue to evolve along that path even after it is built. Consider the construction of a new community almost like the growth of a sapling. Its traits are set, in general. What it can be has been somewhat preordained, but its full potential will not be known until it matures within its environment and bears fruit of its own.

# 10

# THE FUTURE NEIGHBORHOODS

## Phase III, IV, the Commercial Center, and Beyond

> *Can you look your children in the eye and tell them that they can have grandchildren who will have grandchildren? Can you do that? It is about survival of the species. So when our grandchildren are old enough to understand, they should be able to say, "Grandpa did everything he could."*
>
> —C. ALAN NICHOLS, P.E., C.E.M., B.C.F.E., LEED AP

Below is part of an interview between friends Paul Rollins and Wayne Moody.

**Rollins:** What's the one thing you want to say about what's evolved in Civano?

**Moody:** Gosh, it's been six years since I was involved in Civano and 10 or 12 years since the original plan was approved. A lot of time has gone by. And we now have this incredible neighborhood here.

And I went on to do another project after Civano, a much smaller project, which was very hands-on. And I was able to take the experience that I gained with me. I learned a lot from Civano as I'm sure many people did. I was able to take that experience and build on it and actually create something quite unique from that project.

Wayne Moody's community effort to build Milagro Co-housing on the opposite end of Tucson from Civano was a truly pioneering endeavor resulting in an equally revolutionary place. It contains more of its own infrastructure and is based on landscape form and function. The community harvests rainwater into the landscape and hardly sheds a drop into the surrounding desert that has not been slowed, filtered, and released through natural processes—and no more than the property shed before it was developed.

The homes are largely comprised of adobe bricks, a very local and low-embodied energy material which can one day simply become dirt again, if that time comes. The community is quiet and supremely peaceful. There are no cars anywhere near the homes, parks, and sidewalks that connect the small community. One must take a short

**Figure 10.1** Civano community fourth of July parade.

walk to the shaded, gravel parking spots at the end of a long dirt drive with very effective bermed traffic-calming bumps. Or are they devices for shedding rainwater to the shoulder and keeping it from eroding the dirt road? Indeed, Milagro embraces the car, but simply sets it further away.

The community is a few miles away from the nearest services and not in a major commercial growth corridor. So there are ways that this community could be more efficient during the times it was built, but it is built in direct stewardship of the land. Milagro may represent an ideal in fringe development, a community of people who appreciate nature and the opportunity to buy a housing product that allows them to be in a community surrounded by their privately protected community property, the Sonoran Desert. They take care of their water out there, and that serves a benefit to society that is immeasurable—especially in a desert city.

So the pebble moved again when Wayne Moody walked across Tucson and planted a seed of Civano. Milagro is made of the same stuff as Civano, but more keenly refined to a purpose because of Civano. Milagro didn't try to push the mainstream too far. Milagro found its ideal size, purpose, and people—yet Milagro also reinforces the development of the future.

**Moody:** But in looking back at Civano, I still see it as one of the milestones of planning in this era to have accomplished as much as was done—even though I can look back and see all the incredible missed opportunities. I mean, I haven't begun to share some of the ideas and thoughts I think actually could have happened. Such as the botanical gardens and the washes and the job opportunity, where the school is. There're just so many.

And so I look back personally and view it as one of the greatest opportunities I've ever had and one that I can be proud of. It took me a while to get to that point. But I'm proud to know that I had a little influence over this project and it has influenced other projects. And in Tucson, the development standards, the new energy conservation standards, are really quite different now than they used to be before we started Civano. There's a whole new attitude about how you develop a property and so forth.

And so, you know, I think it was worth it from many perspectives.

## FAST FORWARD

The idea that Civano could be completely solar powered is true. In fact, it is truer of Civano than nearly any other master-planned communities of its time because Civano was designed to use less energy to operate in the first place. Therefore, it will take a smaller investment to power Civano with renewable energy resources than a traditional community of the same era. This means those who paid a little bit of extra money to live in an efficient home could potentially reap the full benefits of this advancement first as well.

Maybe one day the regional power grid will be an extremely robust system which includes a very self-sufficient peak load provided by the combined effect of a network of micro-generators. These micro-generators could be anything from a home with a slight net-positive PV system to a farm with a methane generator. The standing supply of factories and other power generators could supply base demand, emergency and backup energy to the grid. Since Civano was built to be more efficient, the neighborhood may have a greater chance of becoming a net provider of electricity. Revenues could be used to fund community activities or systematically remove energy costs from the Civano homeowners.

Of course, these ideas can't be realized today, but they are already on the path to being real today as more people each year retrofit their homes with solar panels, rainwater barrels; and as more ride bicycles to reduce miles driven in hybrid vehicles. If we don't listen to such ideas and place them forward in due time, in a logical succession we can visualize, we won't have a clue which way we're going from step to step.

The idea that Civano could provide one full time job for every other resident is true. The commercial center is not even built yet and there is an entrepreneurial crowd inside of the Solar Village. They bought into a new idea—the innovation of their actions speaking directly to the very spirit of their entrepreneurial efforts. Many living in the village own home businesses, mixed-use properties, and save money on utilities or commuting while making money in their community. The community is operated by people who are interested in what Civano could be, not just what it is or was. They see a living, evolving place with a future created by many visionary stewards.

The idea that Civano could be a car-free or extremely car-light community is true. Not today, but with the Houghton Area Master Plan comes the promise of future development all along the Houghton corridor. The Congress for the New Urbanism is one pattern language which offers guidance for seeing the region as a whole, understanding how HAMP fits within Tucson at large, and how HAMP neighborhoods may be linked to the greater city via public transportation. Will HAMP take on the Civano pattern of development using the new tools at its disposal? The Houghton Area Master Plan: Adopted June 7, 2005, Resolution No. 20101 states that, "The general plan identifies the HAMP area for future growth to be master planned, where the Desert Village model has been established as the future land use pattern."

Civano, however, is not just about the neighborhood form. In fact, some of its greatest successes are its energy and water performance targets and achievements. HAMP master planning to date includes a mandate to "reduce the energy requirements for buildings and transportation throughout the HAMP area, and identify opportunities for renewable and alternative energy resources." These statements are, as yet, undefined and mean no more about performance than the term "green." However, the latest HAMP planning published on the city Web site at the time of this writing indicates thoughts of high performance efficiencies in energy and water consumption similar to those required by the Sustainable Energy Standard. They would surpass current City of Tucson code requirements by pre-defined increments—another green building code language to put into our bag of tools that has already been condoned by the City of Tucson.

The HAMP plan, as published most recently in 2005, would help guarantee a serviceable neighborhood for years to come. They have adopted the language of this new paradigm of design and land planning. Construction processes pioneered by leaders in the industry represented by groups like the Southern Arizona Home Builders Association (SAHBA) and the National Association of Home Builders (NAHB) and the U.S. Green Building Council (USGBC) can be honed and streamlined to provide efficient buildings within regional contexts; with a special connection to the jurisdiction having authority. The City of Tucson utilizes the LEED guidelines combined with the SES protocol for development of its buildings. The University of Arizona has adopted LEED principals. Pima County revised aspects of the NAHB model, LEED model, and other trade influences to create Arizona's first countywide green building program.

The LEED—Neighborhood Development program, once finished, can provide a systematic approach to design processes, charettes and plan development for each neighborhood area within the larger Houghton Area Master Plan because it provides a framework for the processes. The City of Tucson will play a major role in determining which influences are attached to the HAMP development agreement when planning commences. But such decisions will be subject to the public meeting laws of Arizona, and the public will have opportunities to have a voice about how their city evolves, if they so choose to exert it.

Ultimately, all of these aspects would combine with the awaited City of Tucson green building program to make sure that the concept and practices employed in developing HAMP are understood and supported by the jurisdiction having authority. At the end of the day it is our cities that decide from within themselves how the urban landscape will develop. The open meetings which we citizens attend (or do not attend) are the places where these decisions are often played out. While it may seem cumbersome to involve deep public and private partnerships, they have their place in groundbreaking developmental shifts or heavily impactful projects. If Westcor comes back knocking on HAMP's front door, will they choose the path of least resistance and develop to the status quo? Or will they take a lead in reaching out beyond our current hold and find new territory within the possibilities of what we can accomplish at the time, logically, with the tools at hand?

It is likely that the answers to these questions are somewhat predictable by what is said about Civano. Was Civano a success? But success is measured on different scales.

**McDonald:** Is Civano a success?

**Nichols:** Oh, absolutely. It meets all our standards. Okay, understand that the standards that we are talking about are the engineering standards of the project. From the very fundamental beginning of the project, it wasn't about, you know, twisty roads and beautiful buildings, although the village concept was very much the early concept. But it was a solar village, a sustainable village—that was the concept. And if nothing else, we're just a boring load of engineers. We look at the facts, just the facts, only the facts. We have numbers; we crunch the numbers; and we make the rules based on numbers. You know, we don't do subjective things, we do objective things. This is the objective and here's how we do it and here's how we proved it. This is what we do. You know, so all the rest of it, I stayed out of.

I got my little piece. My little piece is the engineering standards. And we picked up water too. Even when I was on the energy commission, we decided we would tackle the water issues of the city because that uses energy. So under the energy commission, we took a lot of liberty on the projects that we chose to pursue. So the water was a 60 percent reduction from potable sources. And so that became a standard that the city imposed that we ended up auditing.

From my standpoint of the project, it is the utility reduction, water reduction standard. As long as we're meeting that, we're a success.

As I've lived here, and absorbed more about what it means to live in this kind of village, urban community, I've learned about the other side of land development. I am saddened because Civano was never intended to be broken into two or more communities. We were always meant to be one community. One thing we might want to do is get it all under one HOA. But I don't know if that can happen. It is something that will have to happen after Pulte has had a fair shake to do its piece and go.

## THE CIVANO INSTITUTE: REVISITED

Civano has proven that if you build it better, they will come—but it doesn't mean they will come for the reasons you originally thought. The building of community is both a simple, instinctual activity, and at the same time a labor much more complicated than master planning of a high performance infrastructure. The community itself is a part of a place, a shared identity and an evolving entity which is only as predictable as human behavior itself. In the first Civano report, as cited in the early pages of this book, the vision was stated:

> "Our goal is to build the BEST COMMUNITY POSSIBLE by merging Southwest traditions with environmental caring, solar design, proven technologies, economic opportunities, social responsiveness and sound financing." Three primary project objectives were put forth:
>
> - ENVIRONMENTAL ACCOUNTABILITY utilizing energy efficiency with solar applications, resource conservation and recycling, biking, and walking for internal transportation, water conservation, and preservation of open spaces.
> - ECONOMIC PERFORMANCE with employment provided for 50 percent of the residents, primarily in environmentally useful occupations in a business district enhanced with home/studio offices and civic/educational facilities.
> - SOCIAL RESPONSIBILITY through affordable, handicapped accessible housing and community child/elder care facilities with continuing educational and youth employment opportunities.

Through the years, the Civano Institute, most recently named the Tucson Institute for Sustainable Communities, has been a flashing beacon—currently off. The organization started off much like the Obama administration has—as a transformational leader sharing a vision and a message about what kind of places we could build to live in, what we could aspire toward. As the vision was accepted and work started, the Institute evolved to support builder education programs, provide a voice through which the developers and the city could speak to the community, and even technical support for planning and development strategies.

The Civano Institute could be revived to provide a conduit for Civano's community visioning; and also to serve as a regional resource to help guide future development throughout the HAMP corridor and if the scale of integration seems appropriate for the times—the county, state and beyond. Civano is not only a resource to those who buy and sell it. Civano and efforts like it are part of the diversity of nature itself—places on the fringe of human adaptation adorned with thoughts about the future. The goals from the first Civano report, the very first widely released with a unified voice, are not all complete. There was no timeframe set—only the goals were set. As Civano moves forward, it will continue to be an example of what can be done, and it will always be compared to what came before and after.

**McDonald:** Do you think that Civano is hurt by having Mesquite Ranch, the mainstream development, so close?

**Nichols:** Well, it's an excellent example of what Civano could have been if we hadn't held the line as much as we could. If you drive through the Civano project and then drive in Mesquite Ranch; you can see that there's a dramatic difference in the way people are building those areas. Theirs is the common way of building with the same model home all the way down the block. One house looks just like another. You know, and that's the architecture. But they also aren't going to do any more than they have to because if they can figure out how to save a hundred bucks, and they're building a hundred homes, they got another $10,000 bucks in their pocket. So it's always natural to look for places to cut the price—and especially ways to cut the price for things you can't see.

You can't see insulation. You can't see glass glazing or insulation on pipes. You can't see roof insulation. You can't see the SEER rating of the air conditioners. Until homeowners get wise, none of that is sellable because you can't see it, which has been the big problem selling the Civano homes. So that's a unique comparison. And it's actually beneficial to us to have that because it's going to continue to be a valuable study to see how the phase 1 homes compare to the phase 2 homes compared to the Mesquite Ranch homes.

In turn, it will be interesting to see how future homes compare to Civano homes if the green building movement flourishes and a group such as the USGBC achieves its mission statement: "To transform the way buildings and communities are designed, built and operated, enabling an environmentally and socially responsible, healthy, and prosperous environment that improves the quality of life." Why? Because the same kinds of pioneering people who built places like Civano have taken those seeds and original ideas to the next evolution. They have not quit and they continue to push the bar on high performance habitat development.

## CAN FUTURE BUILDINGS BE THE ENVIRONMENT?

Arguably, buildings already are part of the environment. The more direct question would be: is there a level of building efficiency which surpasses concerns of consumption? The concept of homes and buildings being micro-generators, producing more energy than they consume, brushes the surface of the idea that a building could provide no less than the energy and water itself requires and do so in a manner which is congruent with nature itself. Homes may one day become a net producer of clean energy and water and their shared common grounds may even become a source of food for the inhabitants living there.

As Civano evolves and more people participate in the farmers' market, maybe another community garden will be created in the village for people to grow produce in—to sell at the farmers' market to other people who live in the village. Maybe as more residents catch on to the idea of collecting rainwater and slowing the rate of runoff from the neighborhood site, the community may become more lush, support more food growth, and be less damaging to downstream biomes. With each step toward the future, we have the opportunity to improve on our living situations by investing in our communities in ways that will help them flourish. One day, maybe the very fabric of our built environment will satisfy much of the urban infrastructure needs we have, for which we currently transform ever larger swaths of earth. Maybe one day, the urban infrastructure will seem almost alive, if only it were not knowingly constructed by people, and provide much of our needs from within.

**McDonald:** Has Civano been an asset to the community?

**Scott:** I would say I think it has. It's a pioneer in its own way. It has spawned many Civanos. There are other little Civanos, not named Civano; but there's a development on the west side now with the same kinds of amenities on the housing in that area. They have solar—Armory Park Del Sol.

Lessons learned have been applied to new places. And frankly, Civano is successful because when they did surveys about Civano, before it started, they wanted to know if there was a market for people to buy into it. And every time they did a survey, and I think there were a few that were done, they had a resounding positive response by the potential new buyers of such houses.

And because of the huge positive response to this—this kind of idea, they just went forward with it. And frankly, Kevin Kelly and David Case assembled an internationally acclaimed group of individuals to help plan it from the ground up, including all of the plants and the way things were done. And I salute them for bringing to the table those with the great expertise to make this thing a practical reality. There were others who were involved with the planning of it. And I think that they together moved forward, as they say, from the philosophical idealistic plane to the practical application of it.

**McDonald:** What do you see Civano's role in the future as being?

**Scott:** I think it has established a standard that people might use in the future. All the lessons learned from what happened there onsite may be implemented and spawn other buildings and, be they commercial, residential, industrial, they will have to take a look and see what happened there. I think Civano can make people stop and think about when they build anything new or retrofit something, how they should go about doing that and what they should put into the equation.

## THE FUTURE

This business of building is about the future. The stones we move today may become a foundation and then, years hence, return to being just a pile of rocks. Everything we move is a piece of the puzzle and every bit of ecosystem turned under the blade holds untold potential. The voices of Civano, far more than those included in this book, have shouted and laughed together, and sprouted ideas together. They talked about what a place could be, if it were really built to be the ultimate home we could collectively imagine. Some ideas were for further in the future, toward which Civano was a manifest stepping stone.

What could the ultimately efficient urban village look like? A Tucson energy engineer once posed some far out ideas back in the early days of Civano's design (Figure 10.2). What he suggested was that maybe…

> The village center could be served from an energy center—a central plant that generates electricity, hot water, and cold water from renewable energy sources.

The central plant could also collect rainwater for use in the cooling towers. The economies associated with large scale central plants would make many alternative sources of energy available to the village center that would otherwise not be cost effective.

In the Sonoran Desert, low wet bulb temperatures combined with the daily temperature dip at about 7 a.m. each day provides an opportunity to store cold water at night for use during the day. Our most abundant energy resource is the large number of days with available direct solar energy. The most direct and most effective solar collectors will be the village structures themselves. The second line of use will be the direct conversion of the sun's energy into useful utilities. Combining the photovoltaic and hot water collection technologies we will collect two types of energy in a small space.

Except for electricity generation, both heating water and cooling water will be generated "off peak." Hot water generated during the day will be stored to provide heat in the early morning and night hours. In contrast, cooling water is best generated at night to provide cooling in the day. Thermal storage will create the advantage of 24-hour operation. While the village center occupancy may vary during the day, the central plant can operate at any time of the day to harvest either heating or cooling water. By using storage the central plant is able to operate at the most optimum times of the day, independent of the occupancy of the village.

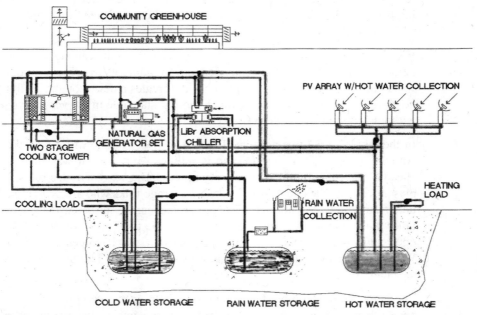

**Figure 10.2** Conceptual drawing of a distributed energy resource facility.

Electricity storage, on the other hand, would not be advisable since most if not all of the electric power generated will be consumed by the village or central plant. Excess electric power, if any, will be sold back to the utility for a profit. The community greenhouse will serve as an integral central plant component.

But he, and others with grand ideas, compromised to get the Civano Memorandum of Understanding, and the associated Sustainable Energy Standard (SES) instead—both printed in the appendix of this book with modifications. These are provided to readers who may want to take their own stand, learn the patterns of a high performance code language, and create a place like Civano. There are other means to a similar end, such as the LEED—Neighborhood Development program, but advice from Inside The Civano Project suggests that strategies should be drawn from each body of knowledge available about high performance building, and utilized on a scale that integrates well with the vision at hand, and the systems through which the goal will come into fruition.

**Nichols:** This project parallels the beginning of the universe. In the beginning, there was a void—a place where nothing had ever happened. There was no time, no space, no dimensions. There was nothing. So in contemplating the evolution of the Universe you have to first try and imagine nothing—nothing at all. Not even a thought.

Then the Universe, in a way, is a thought. The thought is transition from nothing to something. The "something" in our case is the evolution of life, and life's reflection of things; our understanding of things. The Universe began by evolving from nothing to a thought. So, too, the evolution of the Civano project began with a thought.

What we had here was a void, a place where nobody had thought we could go before. Now, it is a wealth of knowledge from attempts to achieve a goal which was unthinkable at first. The Civano project creates a new thought: With little or no extra money, we can become much more sustainable in the very ways we live without sacrificing a modern lifestyle.

In the Universe we live and die. Those who live win and those who die don't. And we are, in fact, living. Civano is selling. We're on our way there. We are ahead of the curve on the evolutionary process. But we're average folks. No hippies ... well, maybe one or two; but we're average folks who are evolving into the understanding that living in the Civano project can be far better than living most other places. The infrastructure can provide the tools we need to build a more open and progressive community. And what if gas becomes $20 per gallon? Would we be able to get food? Would we starve and die? I think our odds are better in a place like this.

The Universe is going on. It is going on to help us evolve into what we need to be. And part of what we need to be is aware—aware that we are in a finite resource pool. But we have had the demonstration that shows us we can, in fact, have a sustainable model where the resources we need to survive are more within ourselves. We are evolving!

It's possible that that evolution will do what it does and Civanoites, or those like us, will make it. But we're not going to make it in a few off grid homes sitting out in the middle of nowhere. We are going to make it together, as a community.

# EPILOGUE

As the research and writing for *Inside the Civano Project* comes to a close, the advancements in Civano are accelerating. The Neighborhood I community center, having been bought from Fannie Mae by Civano resident and community leader, Les Shipley, is about to get its first restaurant.

**Shipley:** If we didn't have this little community meeting space there would be no community. There would be no coffee meetings, there would be no book club, there would be no teen night. There would be nothing. Now there is a meeting space, with a kitchen and food where we can build camaraderie.

At the first-ever Governor's Achievement Award for Innovative School Design, held at the Arizona State Capitol, Tucson's Civano Community School (a.k.a. the Greenest School in America) was awarded top honor in the "Sustainable Design" category. The small school is an example of how a building can inspire students to environmental awareness help teach them to have commitment to increased sustainability and is serving as a model across the state to help inspire new ways of enveloping students in a lifestyle which is more sustainable and making 'green' an everyday part of their lives—not just something they read about.

There are approximately 91 Civano based businesses listed on the Civano website (www.civanoneighbors.com), most of which are home based businesses. Soon, a pub is expected to open in one of the mixed-use buildings surrounding the neighborhood center, to the delight of many of the community's social butterflies. As the community center is coming online, number of job positions in the village will increase as well. The 45-acre, all commercially zoned center for Civano (including Sierra Morado) has yet to even get a foundation put it. When it does, the potential to live and work in 'The Solar Village' will be fully realized on the scale originally visualized.

Outside Civano, large companies like the venerable JC Penney are taking up the cause. Opened under the name, The Golden Rule, Penney is one of several examples of established organizations now striving to follow that moniker by making their buildings more energy efficient and environmentally friendly. The idea that going green saves green is settling in around the country. The seeds of Civano, and people who support ideas like Civano, are making a difference in all areas of our culture, technology and environment.

# GLOSSARY

### Arizona Solar Energy Commission (ASEC) Land Committee

A subcommittee of the Solar Energy Commission (now MEC).

### Barrier to Sustainability

A barrier to sustainability is what it sounds like: a challenge to moving forward with increasingly sustainable practices. Barriers to sustainability can be as obvious as a cultural tendency to prefer SUVs or as subtle as a budgetary practice which creates incentives to "use it or lose it" when saving the resource for next year would have otherwise sufficed.

### Base (Plug or Process) Load

Base loads are those energy demands which are most constant throughout the year. Non base loads are the parts which fluctuate considerably with seasonal changes. Heating, cooling and landscape water use are considered peak demands because they fluctuate.

### Bedroom Communities

A bedroom community has no or few services, to the extent that people who live there mostly need to commute on a daily basis to get to work or other daily necessities. The phrase "bedroom community" comes from the notion that people who live in them basically come home just to sleep, having spent most of their day commuting.

### Biome

A major biotic community characterized by the dominant forms of plant life and the prevailing climate

## Blower-door tests

The blower-door test is a way to measure the leakiness of a building envelope. The technician methodically closes every opening to the home and then installs an enclosure with a fan in one exterior doorway. By measuring the static pressure in the building, the technician can determine how many square inches of opening are in the envelope. This blower door test is one measure of a well-sealed building envelope which is, in turn, a determinant of efficiency.

## Brownfield

The term "brownfield" is used to distinguish these sites from "greenfields," undeveloped land outside of cities and urban areas.

## Candidate Species

Any species that is undergoing a status review that NMFS has announced in a Federal Register notice. Thus, any species being considered by the Department of Commerce Secretary Interior Secretary for listing under the ESA as an endangered or a threatened species, but not yet the subject of a proposed rule (see 50 CFR 424.02). NMFS's candidate species also qualify as species of concern.

## ccf or 'Centum Cubic Feet'

1 ccf = 100 cubic feet = 748 gallons

## Cogeneration

Conventional power plants emit the heat created as a byproduct of electricity generation into the environment. Cogeneration processes capture the byproduct heat for domestic or industrial heating purposes, either very close to the plant, or for distribution through pipes to heat local buildings.

## Congress for the New Urbanism

The Congress for the New Urbanism (CNU) is the leading organization promoting walkable, neighborhood-based development as an alternative to sprawl. For more information visit, www.cnu.org.

## Design Charrette

An intensive design process that involves the collaboration of all project stakeholders at the beginning of a project to develop a comprehensive plan or design.

## Dwight D. Eisenhower National System of Interstate and Defense Highways

The Interstate Highway System is a separate system within the larger National Highway System. The entire system, as of 2004, had a total length of 46,837 miles (75,376 km).

## Endocrine Disruptor

An exogenous substance that changes endocrine function which interferes with hormones and causes adverse effects at the level of the organism, its progeny and/or the population.

## Geographic Information Systems (GIS)

A geographic information system (GIS) captures, stores, analyzes, manages, and presents data that is linked to location. GIS applications are tools that allow users to create interactive queries (user created searches), analyze spatial information, edit data, create maps, and present the results of all these operations.

## Greenwashing

Greenwashing is the inadvertent, or advertent, false representation of a product, process, service, or lifestyle as environmentally friendly. Greenwashing occurs on a spectrum from total falsehood to misnomer.

## HERS Rating

A standardized system for rating the energy efficiency of residential buildings. A HERS rating is only a numerical value and does not define which areas of the home need to be improved or provide recommendations for improvement.

## Heating Degree Days

A measure of how cold a location is over a period of time relative to a base temperature, most commonly specified as 65°F. The measure is computed for each day by subtracting the average of the day's high and low temperatures from the base temperature (65°F), with negative values set equal to zero. Each day's measures are summed to create a heating degree day measure for a specified reference period. Heating degree-days are used in energy analysis as an indicator of space heating energy requirements or use.

## Cooling Degree Days

A measure of how warm a location is over a period of time relative to a base temperature, most commonly specified as 65°F. The measure is computed for each day by subtracting the base temperature (65°) from the average of the day's high and low temperatures, with negative values set equal to zero. Each day's cooling degree days are summed to create a cooling degree day measure for a specified reference period. Cooling degree days are used in energy analysis as an indicator of air conditioning energy requirements or use.

## Houghton Area Master Plan (HAMP)

The HAMP area, which is largely undeveloped south of Irvington Road, offers an opportunity to plan and develop a place where people can enjoy a comfortable environment in which to work, raise children, retire, enjoy being with friends, be close to nature, and pursue a healthy lifestyle.
http://www.tucsonaz.gov/planning/prog_proj/projects/archive/hamp

## Human Scale

The proportional relationship of the physical environment to human dimensions, acceptable to public perception and comprehension in terms of the size, height, bulk, and/or massing of buildings or other features of the built environment. A sidewalk is human scale whereas a street is car scale and a runway is airplane scale.

## Hydrology

Hydrology is the branch of geology that studies water on the earth and in the atmosphere: its distribution, uses and conservation.

## Btu

A British Thermal Unit (Btu) is the amount of heat energy required to raise one pound of water one degree Fahrenheit at one atmosphere pressure.

## Land Selection Committee

John Wesley Miller assembled a land selection committee (chaired by Bob Patrick) to find land to build the Civano project. They looked at land across Arizona and ended up with the parcel on the southeast side of Tucson.

## Leadership in Energy and Environmental Design (LEED)

LEED is a brand created by the U.S. Green Building Council which denotes its green building methodology and approaches to several different development scenarios including but not limited to: new construction, existing buildings, campuses, interiors, and neighborhood development.

## Manual J Calculations

Manual J calculations are the basis for determining correct air conditioning unit sizing. Manual J practices have condoned system over-sizing, which is bad for efficiency. The most recent Manual J, the Eighth Edition, still has been shown to have at least a 10 percent buffer, but this is a drastic improvement over previous years.

## Master Plan

A document that describes, in narrative and with maps, an overall development concept including both present property uses as well as future land development plans. The plan may be prepared by a local government to guide private and public development or by a developer for a specific project.

## Metropolitan Energy Commission: (Social Organization = Law + Processes)

The Tucson-Pima Metropolitan Energy Commission (MEC) was established in 1980 by resolution of the City of Tucson mayor and council and Pima County Board of Supervisors. The goal of MEC is to serve "… as a catalyst for the City of Tucson and Pima County to build a more sustainable energy future in the region." (http://www.tucsonmec.org)

## Mixed-Density

Zoning to allow everything from single family homes with larger yards to multi-family dwellings such as apartments.

## Natural Resources Defense Council

A non-profit environmental action group, the NRDC's mission is to "Safeguard the Earth: Its people, its plants and animals, and the natural systems on which all life depends." For more information about the NRDC, visit: www.nrdc.org.

## One Hundred Year Storm Event

A storm event of such a magnitude unlikely to occur more than once every hundred years on average. These storms are usually described in measurable terms which include a rate, such as: "10 inches of rain in an hour."

## Permaculture

Permaculture is an approach to designing human settlements and perennial agricultural systems that mimic the relationships found in the natural ecologies. It was first developed by Australians Bill Mollison and David Holmgren and their associates during the 1970s in a series of publications. The word permaculture is a portmanteau of permanent agriculture, as well as permanent culture.

Permaculture design principles extend from the position that "The only ethical decision is to take responsibility for our own existence and that of our children" (Mollison, 1990). The intent was that by rapidly training individuals in a core set of design principles, those individuals could design their own environments and build increasingly self-sufficient human settlements—ones that reduce society's reliance on industrial systems of production and distribution that Mollison identified as fundamentally and systematically destroying Earth's ecosystems.

### Planning Permits: (Law)

A planning permit may be issued, depending on the jurisdiction, before the land in question is purchased. In the case of Civano the cost for the planning was reimbursed out of the land auction funds after the plan was approved.

### Rezoning

Rezoning is the modification of the designation of a parcel or group of parcels on the zoning map, which changes the permitted usage of the area. Changes in zoning are usually requested by individuals or businesses and then approved by the zoning commission of a town. Normally these changes are only granted if there is no adverse affect on other properties within the area.

### Riparian

A riparian zone or riparian area is the interface between land and a stream. Plant communities along the river margins are called riparian vegetation, characterized by hydrophilic plants. Riparian zones are significant in ecology, environmental management, and civil engineering because of their role in soil conservation, their biodiversity, and the influence they have on aquatic ecosystems. Riparian zones occur in many forms including grassland, woodland, wetland, or even non-vegetative. In some regions the terms riparian woodland, riparian forest, riparian buffer zone, or riparian strip are used to characterize a riparian zone.

### Social Value Composite Analysis

A system honed by Ian McHarg, the social value composite analysis approach to land planning utilizes map overlays of the region in question, each map with its own social value mapped out (be it forest, lake, river, grazing land or a park). When all these transparencies are layered, one can see the tapestry and densities of social value in the subject area. This type of analysis improves decision making about the impact of proposed development and was a precursor to modern geographic information systems (GIS).

### Solar Energy Systems

Solar energy systems convert solar energy into heat or electricity or harness the light. The "poster child" for solar energy systems are photovoltaic (PV) panels which convert solar radiation into electrical current.

## Solar Parade of Homes

In 1981, 50,000 people attended a tour comprised of solar homes built by seven builders—the early innovators.

## Sonoran Desert

The Sonoran Desert (sometimes called the Gila Desert after the Gila River or the Low Desert in opposition to the higher Mojave Desert) is a North American desert which straddles part of the United States-Mexico border and covers large parts of the U.S. states of Arizona and California and the Mexican states of Sonora and Baja California. It is one of the largest and hottest deserts in North America. The desert contains a variety of unique plants and animals, such as the saguaro cactus. The Sonoran Desert was set aside as the Sonoran Desert National Monument for the purpose of enhancing resource protection.

## Source Energy

Source energy is computed as the energy consumed at the power utility to support the end-use. For every Btu of natural gas delivered, it requires 1.1 Btu of gas energy for dehumidifying it, removing the carbon dioxide, and delivery (approximately). For every kWh of electricity delivered, 3.1 kWs are required for generation losses (approximately 70 percent) and transmission losses (approximately 10 percent–15 percent).

## Section of Land

1 square mile (640 acres).

## The Tucson Solar Energy Commission (MEC)

The Tucson-Pima Metropolitan Energy Commission (MEC) was established in 1980 by resolution of the City of Tucson mayor and council and Pima County Board of Supervisors. The goal of MEC is to serve "…as a catalyst for the City of Tucson and Pima County to build a more sustainable energy future in the region."

MEC is an appointed, all-volunteer civic commission. During the past 23 years, the commission has actively educated the public on a variety of energy subjects, analyzed technical issues, commented on energy legislation, developed strategic plans, issued recommendations to the public, private, and nonprofit sectors, developed community support, and sponsored many energy-related community activities and projects.

MEC led a 12-year participatory planning and design process for Civano.
For more information about MEC, visit: www.tucsonmec.org.

## University of Arizona's Environmental Research Laboratory

From the College of Agriculture's Department of Soil, Water, and Environmental Science: The ERL's mission is to, "Improve the health, welfare, and living standards of communities in desert areas through the application of appropriate and sustainable technologies." For more information about the ERL, visit: http://cals.arizona.edu/swes/erl/index.htm.

### United States Green Building Council (USGBC)

Mission Statement:
"To transform the way buildings and communities are designed, built, and operated, enabling an environmentally and socially responsible, healthy, and prosperous environment that improves the quality of life." For more information about the USGBC or LEED, or to join the organization, visit: www.usgbc.org.

### Urban Sprawl

The compact and mixed-use urban form found before 1950 made for a walkable, efficient urban landscape. Since the widespread adoption of the automobile, separate-use zoning codes and high-volume road standards have made sprawl today's default urban landscape; ironically changing the car from an object of freedom to a vehicle of necessity.

### Variances

Variances are essentially waivers to existing building code or zoning laws which can be sought for building projects. Sometimes special constraints make adhering to the exact letter of the code financially infeasible. These variances are generally made on a case-by-case basis so a review process can ascertain if the project can move on with a variance without being an undue risk to the end users or the builders.

### Vertically Integrated Company

In microeconomics and management, the term vertical integration describes a style of management control. Vertically integrated companies are united through a hierarchy with a common owner. Usually each member of the hierarchy produces a different product or service, and the products combine to satisfy a common need.

### Xeriscape

A landscaping method developed especially for arid and semiarid climates that utilizes water-conserving techniques (such as the use of drought-tolerant plants, mulch, and efficient irrigation).

### Zoning Maps

Zoning maps are maps that are included in the provisions of the Zoning Resolution to indicate the location and boundaries of zoning districts.

# A

# CIVANO IMPACT SYSTEM MEMORANDUM OF UNDERSTANDING ON IMPLEMENTATION AND MONITORING PROCESSES; SIGNED JUNE 26, 1998

The parties to this Memorandum of Agreement are the City of Tucson (the "City") and The Community of Civano, LLC and Case Enterprises Development Corporation (these latter two collectively referred to as the "Community of Civano"). The Community of Civano is the present master developer and the owner of all of the property in the Civano project located east of Houghton Road except for the parcel developed as the Global Solar property as set forth in the Amendment to the Development Agreement which was adopted by the City on June 23, 1997, and property dedicated to the City by the plan for Neighborhood 1 as approved by the City on October 20, 1997. This Memorandum of Understanding is binding upon and shall ensure to the benefit of the Community of Civano and its successors, developers, contractors, builders, and property owners who subsequently acquire rights to develop within the Civano property to the extent applicable to each.

## 1.0 GOAL

The goal of the Civano project is to create a new mixed-use community that attains the highest feasible standards of sustainability, resource conservation, and development of Arizona's most abundant energy resource—solar—so that it becomes an international model for sustainable growth. The State of Arizona through the Department of Commerce, Energy Office has provided significant financial support for the planning and design of Civano. This funding was explicitly in support of the demonstration of the use of solar energy as a guiding, organizing principle of community development.

Another major goal of Civano is to foster creativity and innovation in the construction of Civano. Establishing clear performance achievement levels and then allowing the Community of Civano and designers and builders flexibility in the method of achieving the standards best advances this goal.

The sustainable growth objectives of Civano have been amplified and expanded in the Specific Plan for Neighborhood 1 to include the goals of Building Community, Connection with the Land, Respect for Climate, and Regeneration.

The goal of the Memorandum of Understanding is to confirm the strategies for sustainable development, energy conservation, and economic feasibility of the design and technologies used to implement the IMPACT System which are the basis for Civano and to implement and monitor the Civano IMPACT System ("IMPACT System").

## 2.0 BACKGROUND

The City of Tucson ("City"), in cooperation with the State Land Department and the Metropolitan Energy Commission, and with significant input from the public, established the general development guidelines for Civano. These were implemented through a planning process, the adoption of conditions upon the sale of the state trust land, and conditions enacted in the rezoning of the property by the City. These conditions included resource and energy conservation targets as well as community planning targets such as preservation of open space and encouraging a greater mix of uses.

On October 2, 1995, the City formally adopted the IMPACT System to define and administer the resource conservation goals and to maintain Civano's position on the leading edge of sustainable development. The IMPACT System as adopted clarified the City's initial policies and requirements for the Civano development.

### City of Tucson Guidelines for Developers and Builders: The IMPACT System (Integrated Method of Performance and Cost Tracking)

The IMPACT System is a means of organizing resource efficiency goals and stakeholder cooperation for sustainable community development and for measuring progress toward those goals over time. It is intended to be a cyclical process that:

- Is grounded on metropolitan Tucson baseline conditions that are normally documented and periodically updated by community organizations.
- Is responsive to community policy priorities that will change over time.
- Uses performance targets and specific requirements that exceed baseline conditions without detrimental cost penalties.
- Uses collaboration among stakeholders to reach common goals.
- Measures development performance and costs to evaluate target achievement.
- Enables revisions as baseline conditions improve, and as new targets become technically and economically feasible.

While the initial responsibility for meeting the IMPACT System Standards will lie with the Community of Civano, the responsibility for administering the IMPACT System over time will lie principally with the Civano Community Association (CCA). The CCA is the nonprofit corporation to be created pursuant to the Civano Covenants, Conditions, and Restrictions (CC&Rs) for purposes of administering the CC&Rs, as described in the State Land Department Master Development Plan. The CCA membership will include all owners of Civano property.

The IMPACT System clarifies the Rezoning Conditions established in City of Tucson Zoning Ordinance 7697, and will guide the City's review of subdivision and development plans and initial building permit applications.

The City of Tucson is committed to achieving the original performance targets for Civano over time and does not intend to accept lower performance levels in the initial Memoranda of Understanding.

The IMPACT System established baseline standards and "Performance Targets and Specific Requirements" to achieve the conservation and sustainability goals. These Performance Targets and Specific Requirements are collectively referred to as the "IMPACT System Standards."

The City subsequently entered into the Civano Development Agreement ("Development Agreement") with the Community of Civano, which reaffirmed the goals and principles of Civano and provided the framework for cooperation between the City and the Community of Civano in the development of Civano. The Development Agreement requires both parties to negotiate a Memorandum of Understanding that addresses the implementation and monitoring aspects of the IMPACT System Standards that underpin the Development Agreement. The implementation and monitoring responsibilities described in this Memorandum of Understanding update, clarify, and supplement those in the original IMPACT System.

After purchasing the Civano property, the Community of Civano prepared a specific plan for the development of the first neighborhood of Civano (the "Specific Plan"). The Specific Plan was adopted by the Mayor and Council to further guide the initial development. The Specific Plan, along with the Development Agreement, the rezoning conditions and State Land Patent conditions, provide the framework for developing Civano as a leading sustainable development.

The parties recognize that implementation of the IMPACT System Standards and achievement of the Civano Performance Targets will require a multifaceted approach. While some of the Performance Targets are appropriately addressed by compliance with Specific Requirements for planning, development and construction phases. Other Performance Targets are necessarily dependant upon the actions and conduct of future residents of Civano and must be assessed over time. Initially implementation of the Specific Requirements will involve the review and certification of compliance by the City and by the Community of Civano.

As Civano develops it is anticipated that the parties will have additional information from advances in technology and the monitoring of the implementation of the IMPACT System which may affect future implementation strategies, requirements, and methodologies. It is also anticipated that as Civano develops, the residents will begin to actively participate in the shaping and implementation of the Performance Targets and IMPACT System through their actions and through the Civano Community Association ("CCA").

## 3.0 OUTLINE OF PROCESS

### 3.1 Sustainable Development as a Process

The term "sustainable development" has been defined as "a process of change in which the use of resources, the direction of investment, the orientation of technological development, and institutional change all enhance potential to meet human needs both today and tomorrow."[1]

Conceptually, the IMPACT System Standards are the measuring sticks on which all actions are based and by which performance will be measured. In addition to meeting Specific Requirements and moving toward achievement of all the established Performance Targets, performance must include success in the marketplace. It is understood that the success of Civano will require the good faith commitment and diligent actions by all parties concerned. This approach requires an integrated, flexible response to meet the mandated targets and requirements. This process will be supported and documented by continued monitoring and evaluation of its effectiveness.

### 3.2 Relationship of Plan Review to Monitoring and Evaluation

The underlying principle, as outlined above, is that compliance with the performance targets as described herein will be demonstrated by the construction and subsequent use of site improvements and the development of related programs. Buildings, site improvements, and related programs will be designed to meet or exceed the IMPACT Systems Standards. The Community of Civano and individual builders cannot be sanctioned, however, if actual performance does not meet standards where the personal behavior of occupants affects how buildings function. Plan review will ensure that, in accordance with Section 7, initial development and construction meets Specific Requirements applicable at that stage. Subsequent monitoring of performance in accordance with this Memorandum of Understanding will provide the basis for determining the success in meeting the IMPACT System Standards as well as the basis for improving future conservation and sustainability strategies and standards.

### 3.3 Impact System Evaluation Committee Established

An IMPACT System Evaluation Committee (the "Civano IMPACT Committee") is hereby established to include a minimum of one representative of the Civano principals selected by the Community of Civano; one representative designated by the city manager and one CCA representative selected by the CCA. Commencing January 1, 1999, the Civano IMPACT Committee will meet at least every six months to review the IMPACT System Monitoring Reports to track progress and compliance with the IMPACT System Standards, identify issues of concern, and seek solutions to problems encountered, all in a manner consistent with the success of the Civano Development.

The parties shall further seek the cooperation of the University of Arizona, the State of Arizona Department of Commerce, Energy Office, Tucson Electric Power, other

---

[1] Sustainable Development: A Guide to Our Common Future, United Nations World Commission on Environment and Development. 1987.

utility companies, energy providers, environmental experts, environmental engineers, and similar independent agencies for the monitoring, evaluation and proposed revisions of the baselines and IMPACT System Standards, status of Civano compliance to date, and strategies for improved implementation of the IMPACT System.

## 3.4 The IMPACT System Monitoring Report

The IMPACT System Monitoring Report ("Monitoring Report") shall be a public status report prepared by the CCA and the Community of Civano documenting the success of Civano in achieving the IMPACT System Standards, supplemented with related information from the City and publicly available information concerning the resource conservation baselines and Performance Targets. The Monitoring Reports will initially be submitted to the City at the end of each calendar quarter commencing December 31, 1998 and continuing through December 31, 2000, and shall be submitted thereafter on an annual basis or such periods as may be agreed upon by the parties.

Following is a format for the Monitoring Report which includes not only progress toward meeting the Performance Targets, but also is designed to provide information for public policy changes which would support greater resource conservation and sustainable development. The Monitoring Report will also provide general public information relative to the Civano community development process and progress. It shall include the following information with supporting data submitted by the Community of Civano:

**3.4.1** Overall goals, objectives and strategies, including builder and buyer education for resource conservation practices consistent with the IMPACT System Standards.

**3.4.2** Site layout, showing the built development to date in relation to the overall plans for development.

**3.4.3** Description of anticipated near-term projects.

**3.4.4** Projected timetables and milestones for completion of anticipated projects.

**3.4.5** Status of building and occupancy relative to parking (Parking Inventory and Monitoring Program), pedestrian ways, and landscaping (conceptual planning framework).

**3.4.6** Progress to date, and recommended strategies, toward meeting baseline IMPACT System Standards in the following areas:
- reducing fossil-fuel energy use from gas, electricity and gasoline;
- developing solar energy resources;
- reducing potable water usage;
- reducing building wastes;
- reducing solid wastes;
- reducing auto travel and resultant air pollution;
- creating a balance of jobs and housing;

- creating affordable housing, meeting needs of the onsite working population; and
- demonstrating the economic feasibility of resource conservation and sustainable development as a model for growth in Civano.

**3.4.7** Resulting linkages to central and inner city redevelopment.

The City will provide supplemental information to be attached to the Monitoring Report which includes a status report on progress toward meeting the City's strategies and responsibilities enumerated in this Memorandum of Understanding.

The form of the Monitoring Report will be written, with appropriate graphics, for wide distribution.

## 3.5 Periodic Review

Commencing two (2) months after the initial Monitoring Report and continuing every year thereafter during the life of the Development Agreement as it may be amended or extended, the City and the Civano IMPACT Committee shall consider whether any amendments to this Memorandum of Understanding are necessary to achieve the maximum practical compliance with the IMPACT System Standards. This shall include review of the baselines established in the IMPACT System, the methods of monitoring and establishing compliance, the strategies and Specific Requirements which are appropriate to achieve compliance, the integration of new technology and other matters which are appropriate to maintaining the role of Civano as a leading demonstration of resource conservation and sustainability. Where it is determined that an amendment to this Memorandum of Understanding is appropriate, the parties shall proceed in accordance with Section 8 herein.

## 4.0 JOINT CITY AND COMMUNITY OF CIVANO STRATEGIES AND RESPONSIBILITIES

### 4.1 Description of Joint Strategies and Responsibilities

In recognition of the need for cooperation in achieving the goals described in the IMPACT System, as may be revised from time to time, and this Memorandum of Understanding, the City and the Community of Civano understand that the following strategies and responsibilities are important to the success of Civano as a sustainable community:

**4.1.1** Building plans, development plans, specific plans and similar requests will be prepared, submitted and reviewed in a timely and complete manner.

**4.1.2** The Tucson-Pima County Sustainable Energy Standard will be reviewed and considered by the City for application to other development in the City.

**4.1.3** The parties will explore with the Arizona Department of Commerce, Energy Office and other appropriate agencies, the initiation of programs to further encourage, develop, use, and monitor the beneficial applications of solar energy.

**4.1.4** Grant funding should be pursued to undertake studies and analysis of the role of landscaping and coloration in building and paving cooling strategies.

**4.1.5** Street standards will be reviewed to determine the appropriate methods to significantly reduce the "urban heat island" effect, including the effect of reducing paved surfaces, integrating landscaping for shading of pedestrian, bicycle and automobile parking areas, and allowing alternative, less heat-absorbing paved surfaces in a manner consistent with public safety and handicapped accessibility requirements.

**4.1.6** The parties will cooperate on demonstration projects involving the private and public uses of developing technology; e.g., solar photovoltaic powering of residential homes and of onsite municipal streetlights or pumps, including pursuing grants and other funding for renewable technology demonstration projects.

**4.1.7** The parties will explore the installation of filtration systems for reclaimed water to meet ADEQ standards for full body contact and to permit use of reclaimed water for vegetables to be consumed raw. The parties shall further explore obtaining the adoption of any necessary revisions to existing standards to permit these uses.

**4.1.8** The parties will cooperate to designate and provide garbage and recycling pick-up to all uses, including multifamily and commercial, in conformance with City plans and designs incorporated in the Specific Plan and development plans.

**4.1.9** The City will consider the designation of the proposed Civano recycling center as a "neighborhood recycling center" within the City's overall program.

**4.1.10** The parties will explore the use of Civano as a pilot site for demonstrating new programs and advanced recycling and composting techniques.

**4.1.11** The parties will cooperate to formulate an overall solid waste plan for Civano particularly as it relates to waste reduction and diversion goals.

**4.1.12** The parties will cooperate to develop a transit plan for Civano and the surrounding area, including exploring with the State of Arizona Department of Transportation, the City's Department of Transportation and other affected property owners, a transportation corridor plan for Houghton Road.

**4.1.13** The parties will cooperate with the Greater Tucson Economic Council and other agencies to actively recruit employers for location in Civano.

**4.1.14** The City will consider land-use designations in the vicinity of Civano that incorporate mixed uses, complement Civano, and encourage sustainable development.

**4.1.15** The parties will explore programs to provide assistance to developers/builders who participate in low and moderate-income housing programs.

**4.1.16** The parties will explore the application of the landscape and xeriscape requirements set forth in the City's Land Use Code, Art. III, Div. 7 (originally adopted as Ordinance 7687) to all uses in Civano and to all lot areas consistent with the goals of the Civano Master Development Plan and will further explore the integration of personal and community gardens, orchards and food producing landscaping into these requirements.

**4.1.17** The parties will cooperate to develop a plan for the Pantano Riverpark which integrates adjacent planned open space and recreation areas in Civano.

**4.1.18** The parties will cooperate to develop standards which more effectively utilize natural drainage areas and private open space areas for water harvesting and retention, in order to encourage native plant growth, recharge aquifers and reduce the magnitude of flood flows and erosion hazards.

**4.1.19** The parties will cooperate in a manner consistent with the IMPACT System and the Development Agreement to achieve the goals of this Memorandum of Understanding.

## 5.0 SPECIFIC PROCEDURES FOR IMPLEMENTATION

### 5.1 Developer Requirements Prior to Issuance of Residential Building Permits

Prior to the issuance of any residential building permits, except for proposed model homes for Neighborhood One, Phase One development and permits for the Neighborhood Center in Neighborhood One, the Community of Civano shall submit to the City evidence that the CCA has been established and that the following provisions have been adopted pursuant to the provisions of the Covenants, Conditions and Restrictions which have been recorded and which apply to the development.

**5.1.1** The CCA shall adopt Development Design Guidelines enforceable by the CCA by July 31, 1998. The Design Guidelines shall assure that provisions are made to meet the IMPACT System Standards.

**5.1.2** A Certification Committee shall be formed, and will be a formal part of the Design Review Committee, which is defined and designated in the Covenants, Conditions and Restrictions applicable to Civano and the Design Guidelines. The Certification Committee shall include a Design Review Committee representative, a licensed architect, or a licensed engineer, and a CCA representative.

**5.1.3** Establish exterior water budgets, monitor water consumption, and develop a contingency program to achieve compliance with the budgets if water conservation targets are not met, which utilize City-provided reclaimed water in landscaping for individual residential properties not to exceed 28 gallons per capita per day.

**5.1.4** Establish provisions for utilizing non-potable water for all outdoor irrigation systems and for utilizing efficient and effective, non-pooling drip irrigation systems for all landscaping. This provision shall not require the use of reclaimed water in gardens for the production of raw vegetables for human consumption unless such use is approved by the State of Arizona.

**5.1.5** Establish interior water budgets, monitor water consumption and develop a contingency program to advance compliance with the budgets if water consumption targets are not met, for each building and design the plumbing systems accordingly that will reduce the interior use of water in residential structures to 53 gallons per person per day and to 15 gallons per person per day in non-residential structures. The guidelines shall specify the manner in which water use has been calculated and the principal measures to be taken to meet these budgets.

**5.1.6** The CCA shall adopt and enforce procedures for the review and approval of building plans and energy analyses which demonstrate compliance with the Building Energy Demand reduction required by the IMPACT System. Compliance with the Sustainable Energy Standard, attached as Exhibit 1, as it may be amended with the consent of the parties, which shall not be unreasonably withheld, shall establish compliance with the IMPACT System without further documentation or analysis.

## 5.2 Area Planning, Subdivision and Specific Plan Review

Certain requirements apply to development in a broader context than the specifications for a single structure or lot or a specific point in time. These must be reviewed with consideration for the progress in existing development as well as the prospective development as set forth in the Civano Master Development Plan. City development decisions on the Civano project, which will be viewed in this broader context, are rezonings and specific plans, subdivision plats, development plans, and amendments thereto, not including plans applicable to a single family lot. Interpretation of compliance with these requirements shall be analyzed in the same manner in which the Tucson General Plan, existing area plans and neighborhood plans are applied to prospective development. Any dispute regarding these interpretations may be resolved pursuant to Section 5.4.

General planning areas, not including plans applicable to a single family residential lot, such as development plans, subdivision plats, rezonings and specific plans shall be designed to meet the following IMPACT System Standards, as applicable for the scale of the project, in addition to all other applicable code requirements. The City shall review all such plans in accordance with Section 7. Compliance shall be consistent with the Civano Master Development Plan and may take into account future development as provided for in the Civano Master Development Plan. Monitoring Reports that document compliance with the IMPACT System shall be prima facia evidence of prior compliance. Any dispute regarding these interpretations may be resolved pursuant to Section 5.4.

**5.2.1** Streets and lots are to be designed so that all structures can be oriented to optimize solar exposure and permit the incorporation of some beneficial application of solar energy use in every lot.

**5.2.2** Land may be designated and set aside by the Community of Civano for the design and construction of demonstration projects, as more fully described in Section 6.0, which are compatible with the technologies being offered by Civano.

**5.2.3** Site design and grading plans shall limit site clearance on residential lots to preserve existing desert vegetation and maximize natural drainage in a manner consistent with the grading plans approved with the Civano Master Development Plan and any approved Specific Plan. The exact requirement for preservation of existing desert vegetation shall be determined after completion of drainage and engineering studies.

**5.2.4** Location of a community pool to be constructed early in the development of each phase or neighborhood (as those terms are used in the Master Development Plan) by the Community of Civano and/or individual builders in an attempt to discourage construction of private pools.

**5.2.5** Inclusion of commercial services and other mixed uses with residential developments, consistent with the adopted Master Development Plan and subsequently-adopted Specific Plans, to provide access from residences to commercial or employment areas by walking, bicycling or similar alternatives to automobile use.

**5.2.6** Provision of a central location for access to city bus routes or alternative transit shuttle services, consistent with the adopted Master Development Plan and subsequently adopted specific plans. Consideration of the extension of bus routes to Civano during the development of each Phase or Neighborhood or, as one of several possible alternatives, a shuttle service to the nearest Sun Tran express route and/or park and ride lot sponsored by the Community of Civano.

**5.2.7** Provision of a pedestrian and bicycle-built environment, which is consistent with the requirements to provide access to disabled persons. Each development phase will provide for a majority of through streets (versus cul-de-sacs), construction of a system of sidewalks or bike or multi-purpose paths and nonresidential uses which have orientation access and emphasis on pedestrian /bicycle linkages rather than auto linkages.

**5.2.8** Provide for recyclable materials pick-up areas consistent with requirements of the City Solid Waste Department.

**5.2.9** Provision of a minimum of 300 square feet of non-residential floor area for every two dwelling units constructed. This requirement may be determined by the total development of the Civano provided that the proposed plan does not substantially reduce the total ratio below the minimum requirement. Credit shall be given to dedicated home office space in residential buildings.

**5.2.10** Design of telecommunications capacity to enable the expansion of fiber optics or similar infrastructure to all commercial and home office locations. This may be met by providing conduit capability during construction.

**5.2.11** Identification of the provision for 20% of the dwellings to meet the goal for affordable housing as defined in the IMPACT System Standards to the extent that assistance is available from public agencies, foundations, and other sources to finance and construct affordable housing. The construction of affordable housing shall be reasonably uniform throughout the development of Civano. The Civano development as a whole shall approximately conform with the 20% requirement at the time 625 residential dwelling units have been constructed, at the time 1,250 dwelling units have been constructed, at the time 1,875 dwelling units have been constructed, at the time 2,500 dwelling units have been constructed and at full buildout of Civano.

**5.2.12** Extension by Community of Civano of reclaimed water lines to all lots. Exposed hose bibs for reclaimed water shall be discouraged and if used shall be clearly identified prior to sale of the property.

**5.2.13** Functioning systems using reclaimed water, graywater or rainwater harvesting shall be provided for all landscape irrigation except that:

**5.2.13.1** The limited use of potable water for personal and community gardens producing vegetables to be consumed raw will be allowed within the overall landscape design until it is determined that such use of non-potable water is safe.

**5.2.13.2** Potable water may be used for temporary and periodic flushing of the reclaimed system if necessary to assure efficient operation of irrigation systems, upon notification to and consent by the City, which consent shall not be unreasonably withheld.

**5.2.14** All landscaping which is required by Specific Plan 6A for public streets adjacent to residential properties shall be limited to drought tolerant plants as established by City Development.

## 5.3 Building Plan Requirements and Review

All plan submittals for building permits shall be determined in accordance with Section 7 to meet the following requirements in addition to all other applicable codes.

**5.3.1** All building permit applications shall be certified in accordance with Section 7.0 as complying with the following:

**5.3.1.1** Residential building plans shall provide a certification that the plans as submitted provide for a total energy use through the building shell, heating and cooling systems ("building energy use") of at least a 65% reduction for each dwelling from the 1990 Metropolitan Energy Commission annual energy use baseline commencing at the time of initial residential occupancy. The certification shall be in the form attached hereto as Exhibit 2.

**5.3.1.2** Non-residential building permits shall provide a certification that the plans as submitted provide for a total energy use through the building shell, heating and cooling systems ("building energy use") of at least a 55% reduction for each structure from the annual energy use by a comparable non-residential structure in 1990 as established by the Metropolitan Energy Commission. The certification shall identify the 1990 level used, the method of determining that level and the source material documenting that level. The energy conservation shall commence at the time of initial occupancy. The certification shall be in the form attached hereto as Exhibit 2.

**5.3.1.3** Building plans shall identify the manner in which the proposed structures will be designed to optimize solar orientation for passive heating and cooling purposes, consistent with Civano's goals.

**5.3.1.4** Plans shall incorporate some beneficial use of solar energy to reduce the energy demands from heating, cooling and interior water heating. Solar devices such as currently found in A.R.S. § 44-1761 shall qualify as beneficial uses of solar energy and will satisfy this requirement.

**5.3.1.5** Landscape and hardscape coloration and/or vegetation shall be used to reduce the microclimate temperature adjacent to the structures. The average reflectivity of all major landscape and hardscape surfaces must be 0.5 or greater on the albedo scale or result in equivalent energy savings.

**5.3.1.6** Plans shall identify procedures for preserving construction materials for recycling during construction and for the use of recycled construction materials in construction.

**5.3.2** Structural calculations demonstrating that the roof will support solar photovoltaic, solar thermal power generation and solar water heating systems of sufficient size for the potential uses of the building.

**5.3.3** Location and installation of plumbing stubouts for solar hot water heaters shall be required and shown on all residential and commercial buildings.

**5.3.4** Two water supply systems shall be shown: one for potable water and one for reclaimed water for landscaping or similar external uses. Provisions may be made for rainwater harvesting and/or graywater use for landscaping in conformance with existing codes.

**5.3.5** Non-residential space conditioning system cooling towers rated at one hundred tons or more of cooling capacity shall comply with ADWR water conservation requirements.

**5.3.6** Solar thermal water heaters or other devices or technologies which achieve equivalent energy savings in the heating of hot water shall be included on all model homes for demonstration purposes and as options on all other homes.

**5.3.7** Plans shall provide for built-in recyclable separation features and storage of hazardous materials.

**5.3.8** Provision of electric cart charging facilities, which may include designated electrical outlets which are accessible for electric cart charging.

## 5.4 Expedited Review of Interpretations

The City will establish a review committee for the resolution of all interpretive or technical disputes in accordance with Section 6.2.1 of the Development Agreement (the "Interpretive Review Committee"). The Community of Civano or the City may submit any matter to this Interpretive Review Committee for a determination as to whether the matter is one of interpretation or whether there exists an established requirement which is subject to an established administrative appeal procedure. Where the matter is one of interpretation, the Interpretive Review Committee shall determine the interpretation to be applied. The Interpretive Review Committee shall consist of the Director of Special Projects for the City, a designee of the City Attorney's office and a designee of the director of the department or departments which is/are responsible for the review and/or enforcement of the matter being submitted. The Interpretive Review Committee shall reach a decision on the matter or shall state the reason why a decision cannot be made within five working days of the submission. The Community of Civano may appeal (a) any such decision or (b) the Interpretive Review Committee's failure to reach a decision within five working days to the City Manager pursuant to Section 6 of the Development Agreement. The Community of Civano agrees to comply with any decision that is not appealed to the City Manager within five working days of the decision.

## 6.0 DEMONSTRATION PROJECTS

The parties recognize that development of innovative designs and technologies for resource conservation and use of solar energy are important goals of Civano. In order to encourage such innovation, the Community of Civano may set aside a limited number of residential or commercial lots for construction of demonstration buildings.

Such buildings shall not be subject to the requirements of Sections 5.2 and 5.3 at the time of initial planning or permit review but shall provide descriptions of the manner in which these requirements will be met. Such designated structures shall be reviewed after one year for compliance with the resource conservation requirements of the IMPACT System. The buildings shall not be sold or otherwise conveyed to private parties other than the Community of Civano, unless such buildings are certified as set forth in section 7 to comply with the resource conservation requirements of the IMPACT System Standards and the requirements of Section 5.3. Such demonstration buildings may be leased or otherwise occupied without certification of compliance. If in compliance with the requirements of Section 5.3, the demonstration projects may be marketed to private parties. The Community of Civano shall provide a notice that a building was constructed pursuant to this section to any prospective user or purchaser prior to the use, lease or sale of the structure to the public.

## 7.0 CERTIFICATION OF COMPLIANCE

The parties recognize that the IMPACT System includes both Performance Targets to be reached over time, as provided in Sections 3 and 4, and specific resource conservation requirements, as provided in Section 5, which establish minimum thresholds for performance and which are to be met commencing with the initial development of Civano in order to establish progress toward achieving the Performance Targets. The Community of Civano agrees to establish this compliance through the Monitoring Report and compliance with Specific Requirements as set forth below.

**7.1** During the initial development of Phase One of Civano, and thereafter during the development of Civano until changed by mutual agreement of the parties, the Community of Civano shall provide to the City a certification based upon information provided to it by a professional chosen by the Community of Civano that the plans subject to the certification are in compliance with the conservation requirements set forth in Section 5.3.1. The form of the certification is attached as Exhibit A and made part of this MOU.

**7.1.1** The City may audit any such certification and may request in writing supporting documentation from the Community of Civano . The Community of Civano shall provide such documentation within fifteen (15) days of receipt of the notice. If it is determined by the City based upon such an audit that the conservation requirements of Section 5.3.1 have not been met for one or more buildings, the City shall notify the Community of Civano and the applicant submitting the building plan(s) in writing of the specific matters which are not in compliance, the "Noncompliance Notice."

**7.1.2** The Community of Civano shall have thirty (30) days from the date of the notice, unless the time is extended in writing by the City, to either cure the noncompliance or to submit a plan to correct the noncompliance, the "Cure Statement," which is acceptable to the City. The City shall have fifteen (15) days from receipt of the Cure Statement to either accept or reject the Cure Statement as submitted or to request further information or actions. The parties may thereafter continue to seek a mutual resolution of the problem.

**7.1.3** If the City rejects the cure as proposed by the Community of Civano pursuant to section 7.1.2, the Community of Civano may submit to the City a supplemental plan for curing the non-compliance, the "Supplemental Compliance Plan." If a Supplemental Compliance Plan is submitted, no further action shall be taken regarding the non-compliance for at least forty-five (45) days. By submitting the Supplemental Compliance Plan, the Community of Civano agrees that all requests for permits which contain a substantially similar non-compliance problem will be put on hold pending final resolution of the issue.

**7.1.4** If the City and the Community of Civano are unable to resolve differences regarding the Noncompliance Notice and the Cure Statement within sixty (60) days of the date of the Noncompliance Notice, or differences regarding the Supplemental Compliance Plan within forty-five (45) days of the submittal of the Plan, the matter may be submitted by the City or by the Community of Civano to the City Manager for final resolution.

**7.1.5** A City audit of a plan shall not delay the processing or approval of the plan.

**7.1.6** Upon issuance of the Noncompliance Notice, the City may delay issuance of permits for the plan subject to the notice and all other plans which contain the same potential noncompliance problem until the question of compliance with the requirements of Section 5.3.1 is resolved. The City shall include notice that issuance of permits will be delayed in the Noncompliance Notice.

**7.1.7** If the parties agree that a plan is not in compliance pursuant to sections 7.1.1 and 7.1.2 or the City Manager determines that the plan is not in compliance pursuant to Section 7.1.3, the City may deny the issuance of permits to the proposed structure and any other structures which would not be in compliance for the same reason. In determining whether to deny the issuance of permits the City shall consider the materiality of the noncompliance, the Community of Civano's ability to correct the problem with respect to future buildings, the number of building which would not be in compliance if permits were issued, the cost of curing the noncompliance in the proposed plan and the financial cost to the builder or developer which would result from any denial of permits.

**7.1.8** If the City determines that the certification pursuant to Section 7.1 has resulted in a material noncompliance with the requirements of Section 5.3.1, the City may require that future review and approval of compliance with one or more of the requirements in Section 5.3.1 shall be determined by the City Development Services Department rather than by the Community of Civano.

**7.2** During the initial development of a Phase or Neighborhood of Civano, and thereafter during the development of Civano until changed by mutual agreement of the parties, the City Development Services Department shall, pursuant to its standard procedures including the availability of independent third party review as provided in the 1994 Uniform Administrative Code, Section 103, where appropriate, determine compliance with all Specific Requirements set forth in Sections 5.2 and 5.3 other than the requirements of Section 5.3.1, subject to review by the Interpretive Review Committee and appeal to the City Manager as provided in Section 5.4 of this Memorandum of Understanding and Section 6.2.1 of the Development Agreement.

**7.3** The parties recognize that the goal of Civano is to integrate energy and resource conservation principles, standards and technologies into the standard practices and procedures of the City. Thus the parties anticipate that as experience is gained over time with the implementation of the IMPACT System, the parties will be able to develop less burdensome compliance review procedures without any reduction in the progress toward achieving the Performance Targets.

## 8.0 AMENDMENT

The parties may periodically review this Memorandum of Understanding to ensure that it continues to promote the energy and resource conservation and sustainable development goals of Civano and may amend this Memorandum of Understanding by mutual agreement or as set forth herein to achieve the Performance Targets and Specific Requirements of the IMPACT System Standards and to meet changing circumstances as Civano development proceeds.

If, on the basis of the Monitoring Reports, the analysis and recommendations of the Civano IMPACT Committee, or independent information which has been reviewed by the Civano IMPACT Committee, either party determines that development is not progressing toward full compliance with the IMPACT System Standards in a satisfactory manner, it may notify the other party in writing of its intent to amend this Memorandum of Understanding and/or the IMPACT System. The City of Tucson shall provide a copy of the notice to the Arizona State Department of Commerce, Energy Office. The parties shall then negotiate in good faith to establish appropriate amendments to ensure compliance with the goals and requirements of the IMPACT System. Any amendment shall consider the economic impact of the proposed requirement upon the development of Civano and the investment of public funds and grants in this development. If the parties are unable to mutually agree upon amendments within sixty (60) days, they shall discuss any issues with the City Manager, and the City Manager may, as provided in Section 6.2.2 of the Development Agreement, resolve such issues and impose amendments which he deems reasonably necessary to achieve the goals of the IMPACT System.

Any amendment to this Memorandum of Understanding shall not apply to any application for development approval, including specific plans and rezonings, subdivision plats, development plans and building permits, that has already been submitted for review to the City prior to the notice of the intent to amend this Memorandum of Understanding, unless such development approval expires pursuant to existing code provisions. The parties shall provide notice of any amendment to the Arizona Department of Commerce, Energy Office. Any such amendment shall be in writing. Amendments to the IMPACT System Standards shall be approved by the Mayor and Council.

Any application by Civano for a change of the Civano Master Development Plan, the rezoning conditions or the conditions of any adopted specific plan which are adopted by the Mayor and Council subsequent to the signing of this Memorandum of Understanding shall supercede any requirement herein.

## 9.0 REMEDIES

The IMPACT System Monitoring Report, Periodic Evaluation and Specific Procedures for Implementation set forth in Sections 3.2, 3.3 and 5 above shall be the only portion of the Memorandum of Understanding which shall be subject to the remedies provided in Section 9.10 of the Civano Development Agreement as amended. In addition to these remedies, the City shall not be required to issue any building permit which does not conform to City codes, existing and future specific plans and zoning and the requirements which are included in Section 5, Specific Procedures for Implementation, as set forth herein or as amended or revised pursuant to Section 8.

## 10.0 NON-WAIVER OF COMPLIANCE

Except as may be expressly agreed in writing, any decision by the City approving further development without complete compliance with all requirements and targets shall not constitute a waiver of any future application of requirements or Performance Targets as set forth in this Memorandum of Understanding or in the IMPACT System Standards.

Signed this ____ day of _____, 1998.

_____
Authorized Representative
The Community of Civano, LLC

_____                    _____
Case Development Enterprises Corporation     City Manager
City of Tucson

**268    APPENDIX A**

## MOU EXHIBIT 1

### Sustainable Energy Standard

4/22/98

The following modifications to the CABO Model Energy Code, 1995 Edition are deemed to be a sustainable standard:

*(Editorial Note: This Energy Standard was reviewed by the Tucson/Pima County Building Code Committee and is regionally specific to the Tucson Metro area.)*

Material to be added is shown in *italics*. Material to be deleted is shown as ~~strikeout~~.

## CHAPTER 1

### Administration and Enforcement

**Section 101.4 Scope. Add a paragraph to read:**

*The calculated Target annual energy consumption of the building shell and mechanical system and domestic hot water heating shall be less than the energy required by the present Tucson/Pima County Model Energy Code by 50 percent.*

**Section 102.1.3. Add to first paragraph:** *and there shall be a verification of proper installation before drywall installation and the completion of the "Insulation Installation Warranty" and signature by a representative of the developer and builder.*

**Section 102.2. Maintenance Information:** *Delete the first two sentences.*

**Section 102.3. Change the first sentence to read:** *Whole-window assembly U-factor, solar heat gain coefficient, visible light transmittance and air leakage values of fenestration products ...*

**Table 102.3a. Add the following notes at the end of the table:**

*The minimum design characteristics to qualify as a Thermal Break are:*

   a. *The material used as the thermal break must have a thermal conductivity of not more than 3.6 Btu/inch/hr/sq. ft./F, and;*

   b. *The thermal break must produce a gap not less than 0.210 inches, and;*

   c. *All metal members of the product exposed to interior and exterior air must incorporate a thermal break meeting the criteria in (a) and (b) above.*

*In addition, the product must be clearly labeled by the manufacturer that it qualifies as a thermally broken product. Non-metal products may include metal fasteners, hardware, and door thresholds.*

*For all dual glazed products, adjust the listed U-values as follows:*

   a. *Subtract 0.05 for spacers 7/16 inches or wider.*

b. *Add 0.05 for products with dividers between panes if the spacer is less than 7/16 inches wide.*

c. *Add 0.05 to any product with true divided lites (dividers through the panes).*

**Section 102.4 Equipment. Add a new subsection to read:**

**102.4 Equipment.** *Residential buildings constructed under the provisions of this standard shall be permitted to use refrigerated air conditioning systems selected under the guidelines of the Air Conditioning Contractors of America (ACCA) Manual J Procedures, Specifically Sections 7-27, 7-28 and 7-29 at outside conditions of 105 degrees F. and inside conditions of 75 degrees F. Other provisions of this standard notwithstanding, air conditioning equipment shall have a minimum SEER of 12 or a minimum EER of 10.*

*Evaporative cooling is encouraged for cooling or to reduce air conditioning requirements but may not be used as the method of compliance to this standard except for commercial buildings that use evaporative cooling as an economizer cycle on a refrigeration or air conditioning application. Duct leakage through the evaporative device must be minimized during air conditioning and heating modes of operation. Separate duct systems or whole house ductless ventilation is recommended.*

*Examples of water heating systems demonstrating compliance are listed here:*

– *Solar water heaters.*

– *Instant gas water heaters with electronic ignition.*

– *Heat pump electric water heaters.*

– *Heat recovery water heaters from air conditioning or other sources.*

– *Gas water heaters exceeding 90% efficiency (condensing types) .*

*Commercial buildings or domestic water heating systems which serve only hand sinks, a single mop sink or other applications which have low hot water demands on an annual basis may use any of the following:*

– *Instant electric water heaters.*

– *Point of use electric water heaters.*

– *Storage tank electric water heaters not exceeding 20 gallons in volume.*

**Section 104.1 General:** Delete parenthesis around last sentence, delete the footnote.

Add a sentence after the exception to read:

*Plans and specifications shall show the method of utilizing "beneficial use of solar energy."*

## CHAPTER 2

## Definitions

**Section 201 Definitions:** Revise as follows:

**201.1 Application of Terms.** Conditioned Floor Area:

Delete the words "The horizontal projection of," and capitalize the new first word "That."

**201.1 Application of Terms.** Positive Cooling Supply:

Insert *including evaporative cooling systems,* between "cooling" and "deliberately."

Add a new subsection:

*201.1 Application of Terms. Civano: A Tucson Solar Village, a model sustainable community; a vision of the future where resource consumption is reduced through more efficient technologies, use of solar energy and lifestyles which promote greater harmony and balance with the natural environment; a community in the spirit of the "Civano" period, a golden era of the Hohokam culture which balanced natural resources and human needs; incorporates and demonstrates strategies for achieving more sustainable development.*

*201.1 Application of Terms. Sustainable Development: "Development that meets the needs of the present without compromising the ability of future generations to meet their needs." (UN World Commission on the Environment and Development)*

*201.1 Application of Terms. Beneficial Use of Solar Energy: The following devices/methods may be used to demonstrate compliance:*

- *Solar thermal or solar electric space heating systems.*
- *Trombe wall or clear view collectors for space heating.*
- *Solar photovoltaic systems.*
- *Solar thermal/electric power generating systems, including stand-alone and grid connected parabolic trough and dish stirling.*
- *Solar daylighting systems specifically designed to capture and redirect visible solar energy while controlling infrared energy (conventional skylights are specifically excluded) for at least one half of the non-bedroom space.*
- *Passive building heating for the winter through the use of optimum window shade structures and orientation.*
- *Solar water systems for domestic water heating or space heating.*
- *Solar pool or spa water heating.*
- *Solar oven that is built into the structure.*
- *Solar food dehydrator that is built into the structure.*
- *Solar water distiller attached to building.*

***201.1 Application of Terms.*** *Power Density: The total connected power load of all components of a building system, including all auxiliary components and circuitry, without regard to the timing, scheduling, or control of their operation, in w/ft² or Btuh/ft².*

***201.1 Application of Terms.*** *Site Energy: Energy, other than recovered energy, utilized for any purpose on the site.*

Source energy consumption shall be determined by multiplying the site energy usage in kBtuh per square foot by the following factors:

| Site Energy | Factor |
|---|---|
| Electric | 3.10 |
| Gas | 1.11 |
| Wood | 1.00 |
| Solar (amount of displaced electric or gas) | 0.00 |

**201.1 Application of Terms.** SOLAR ENERGY SOURCE. Revise to read:

*SOLAR ENERGY SOURCE. Natural daylighting or thermal, chemical or electrical energy derived directly from conversion of incident solar radiation.*

**201.1 Application of Terms.** Water-chilling Package of Absorption. Revise to read:

Water-chilling package, absorption.

## CHAPTER 3

### Design Conditions

**Table 302.1** Exterior design conditions: Revise the table as follows:

Table 302.1
Exterior Design Conditions[3]

| | | |
|---|---|---|
| WINTER | DESIGN DRY BULB TEMP. | *30 F* |
| SUMMER | DESIGN DRY BULB TEMP. | *104 F* |
| | DESIGN WET BULB TEMP. | *66 F* |
| DEGREE DAYS HEATING | | *7000* |
| DEGREES NORTH LATITUDE | | *32* |

[3] This table is not intended to be used for the purpose of system or equipment sizing.

Add a new section to read:

## SECTION 304
## WOOD-BURNING STOVES AND FIREPLACES

**304.1 General.** *A wood-burning stove or fireplace shall be considered as providing the required space heating energy only when installed as backup energy for a solar-thermal collection system.*

**304.2 Wood-burning stoves.** *Wood-burning stoves shall be labeled to show compliance with the following U.S. Environmental Protection Agency (EPA) standards for particulate emissions during operation:*

Stoves with catalytic elements              4.1 grams per hour

Stoves without catalytic elements           7.5 grams per hour

*Catalytic stoves shall have an accessible, modular, replaceable catalyst element.*

**304.3 Fireplaces.** *Wood-burning fireplaces shall produce useful heat and be provided with a means of supplying 100% of the combustion air for operation from the outside, and shall limit particulate emissions to less than 7.5 grams per hour. All fireplaces shall be provided with a tight fitting glass door and a positive means of circulating the heated air in the occupied space.*

*Direct vent gas fireplaces shall have a minimum of 70% overall efficiency.*

# CHAPTER 4
# Residential Building Design By Systems Analysis and Design of Buildings Utilizing Renewable Energy Sources

**Section 402.1 Energy Analysis.** Change the first paragraph to read:

**402.1 Energy Analysis.** Compliance with this chapter will require an analysis of the annual source energy usage as required in section 101.4, hereinafter called an annual energy analysis *or shall not exceed the source energy usage shown in Table 402.1.*

Table 402.1

| Building | kBtu/sq. ft./yr. (source consumption) | | |
|---|---|---|---|
| Sq. Ft. Range | Heating | Cooling | Total |
| <1000 | 5 | 22 | 27 |
| 1000–1399 | 4 | 18 | 22 |
| 1400–1799 | 4 | 16 | 20 |
| 1800–2199 | 4 | 15 | 19 |
| >2200 | 4 | 14 | 18 |

**402.1.1 Input Values for Group R buildings.** Add a sentence at the end of the first paragraph to read:

*Domestic hot water energy use must be calculated separately from glazing systems, heat storage, thermal envelope and space conditioning equipment and must meet the energy reduction percentages of Section 101.4.*

Distribution System Loss Factor on page 12; change the Outside factor for Cooling to read *0.75.*

Add the following:

***402.1.1.1 Thermal Mass.*** *Designs utilizing thermal mass may be used provided the materials' volumetric heat capacity is between 18 minimum to 30 Btu/cu.ft. F (except water walls) and meet the values shown in Table 402.1., with walls without external insulation at least 12 inches minimum thickness or 8 hours time lag. External insulation can be used (R-9 to R-11) to reduce thickness of thermal mass to no less than 4". Surface area of uncovered thermal mass (in the direct sun zone) will be minimum 9 times the area of south glass, with $1ft^2$ of additional south glass for every $40 ft^2$ of mass located outside the direct sun zone (a simplified method of calculating thermal mass and south glass areas).*

*Table 402.1*
*Heat Storage Properties of Materials*

| Material | Specific Heat (Btu/lb F) | Volumetric Density (lb/cu.ft.) | Heat Capacity (Btu/cu.ft.-F) |
|---|---|---|---|
| Poured Concrete | 0.16–0.20 | 120–150 | 19.0–30.0 |
| Clay Masonry | 0.19–0.21 | | |
| Molded Brick | | 120–130 | 19.0–27.3 |
| Extruded Brick | | 125–135 | 23.8–28.4 |
| Adobe | 0.20–0.24 | 80–106 | 16.0–25.4 |
| Concrete Masonry | 0.19–0.22 | | |
| CMU | | 80–140 | 15.2–30.8 |
| Brick | | 115–140 | 21.9–30.8 |
| Pavers | | 130–150 | 24.7–33.0 |
| Water | 1.0 | 62.4 | 62.4 |

*402.1.1.2 Summer Ventilation.* Thermal-mass buildings shall be provided with a means of venting to the outside at night during the months of May through October to avoid overheating. Operable windows totaling at least 20 percent of the total glazing area, located for effective cross-ventilation or ceiling fans or a whole-house fan sized to provide 10 air changes per hour may be used.

**Section 402.4 Calculation Procedure. Operational Characteristics**: Add a sentence to read:

*The occupied mode shall be not less than 10 hours in a 24-hour period.*

**Section 402.5 Documentation.** Delete the exception.

## CHAPTER 5

## Residential Building Design by Component Performance Approach

**Section 502.1 General.** Add:

*Thermal design parameters to be used in this chapter are found in Chapter 3, Section 302.1.*

**Section 502.1.2.** Revise the second paragraph by placing a comma after "Masonry," and adding "*earthen materials,*" between "Masonry" and "or."

**Table No. 502.1.2c.** Delete the phrase, "~~SUCH AS A LOGWALL~~" in the heading.

**Section 502.2.1 Walls.** Add a subsection to read:

*502.2.1.2 Glazing.* All glazing facing between 20–165 degrees or 195–340 degrees shall have a minimum summer shade or shading coefficient of 0.39. All glazing facing between 165–195 degrees shall have a minimum summer shade or shading coefficient of 0.5 or less. This may be accomplished by the use of overhangs, covered porches, tinted glazing, or other approved methods.

**Table 502.2.1a** Revise as follows:

Table 502.2.1a[1]
HEATING AND COOLING CRITERIA

| ELEMENT | MODE | TYPE A-1 BUILDINGS $U_o$ | TYPE A-2 BUILDINGS $U_o$ |
|---|---|---|---|
| Walls | Heating or cooling | *0.11* | 0.17 |
| Roof/ceiling | Heating or cooling | *0.026* | 0.026 |
| Floors over unheated spaces | Heating or cooling | 0.05 | 0.05 |
| Heated slab on grade | Heating | R-Value *8* | R-Value 8 |
| Unheated slab on grade | Heating | R-Value *0* | R-Value 0 |
| Basement wall[2] | Heating or cooling | U-Value *0.095* | U-Value 0.095 |
| Crawl space wall[2,3] | Heating or cooling | U-Value *0.06* | U-Value 0.06 |

[1] Values are determined by using the graphs (Figure Nos. 1, 2, 3, 4, 5 and 6) contained in Chapter 8 using heating degree days as specified in Section 302.

[2] Basement and crawl space wall U values shall be based on the wall components and surface air films. Adjacent soil shall not be considered in the determination of the U value.

[3] Typical foundation wall insulation techniques can be found in Standard RS-20 listed in Chapter 8.

**Add a new section 502.2.1.1 to read:**

*502.2.1.1 Software. Model Energy Code (MEC) software (DOS version) called MEC check is available to verify compliance with this standard. It is available for download free from the Internet at:*

http://www.energycodes.org/meccheck/mecdownload.html

*To use the software:*

*Use DOS to modify the LOCATION file line 188 to read: Civano Tucson, 7000, 34892*

*Save the file and open the program.*

*Under state, select Arizona and under city, select Civano Tucson.*

*Select "Trade-offs" from the top bar and set HVAC Efficiencies for an air conditioner value of SEER 12.0 or better and Furnace AFUE to 80 or better. Corresponding heat pump values may be used.*

*Enter the building components and verify the building meets the MEC check criteria.*

*Print the report with the checklist (see Table 502.2.1.1 below) and submit for a permit.*

*Table 502.2.1.1*

*Checklist to Accompany MEC Check Report when using Chapter 5*

| Section | Description | Check |
|---|---|---|
| 502.2.1.2 | Glazing facing between 195–340 and 20–165 degrees has a summer shade or shading coefficient of 0.39 or less. Glazing facing between 165–195 degrees has a summer shade or shading coefficient of 0.5 or less. | |
| 502.3 | Air leakage warranty verifying maximum of 0.35 ACH. | |
| 503.2.4 | Recessed lighting fixtures when installed in building envelope is constructed to accept only lamps with efficacy greater than 40 lumens/watt. | |
| 503.8.1 | Duct Leakage Test passed before drywall and "Intermediate Verification and Warranty" form is signed. | |
| 101.4, 102.4 | Water heater system demonstrating compliance. | |
| 504.2.1.1 | Electric storage water heaters has a standby loss not to exceed 4 watts/ft$^2$ of tank surface or 43 watts, whichever is greater. | |
| 504.5.1 | Pools and spas utilize solar energy as the only water heating source. | |
| 504.5.4 | Recirculating system (if used) is installed on hot water line with a timer and pipe insulation. | |
| 504.8 | Low water use plumbing fixtures. | |

***Section 502.2.4.*** *Slab-on-grade floors: Delete in its entirety.*

***Section 502.3.*** *add the following sentence at the end:*

*An Air Leakage Warranty verifying a maximum of 0.35 ACH shall be provided to the home owner. A representative of the developer and/or builder will perform a blower door test after completion but before occupation of the residence. The representative will certify a maximum of 0.35 ACH based upon the results of the blower door test. An Air Leakage Warranty verifying a maximum of 0.35 ACH shall be provided to the homeowner.*

**Section 502.3.4. Recessed lighting fixtures.** Revise the first paragraph to read:

**503.2.4. Recessed lighting fixtures.** When installed in the building envelope, recessed lighting fixtures shall be constructed so as to *accept only lamps with efficacy greater than 40 lumens/watt, and* meet one of the following requirements:

**Section 503.2.1. Calculation procedures.** Add a sentence to read:

*Equipment not covered by the tables in this section shall show the following maximum installed cooling power densities, including all auxiliaries:*

*Less than 65,000 BTU/hr:*

*2.7 kW primary energy per 1000 sf at site*

*65,000–135,000 Btu/hr: 3.6 kW per 1000 sf*

*135,000–250,000 Btu/hr: 3.7 kW per 1000 sf*

*greater than 250,000 Btu/hr: 2.0 kW per 1000 sf*

*with energy intensity of mechanical cooling equipment calculated from EER ratings if entirely electric.*

**Table 503.35a.** Change the SEER values in the fourth column (Minimum Performance) from ~~10~~ and ~~9.7~~ to *12.0* and *10*, respectively. *Add the following note at the bottom of the table:*

*Note: Air-conditioning may be used selected under the guidelines of the Air Conditioning Contractors of America (ACCA) Manual J Procedures, Specifically Sections 7-27, 7-28 and 7-29 at outside conditions of 105 degrees F. and inside conditions of 75 degrees F.*

**Section 503.4. Add new sub-sections to read:**

*503.4.3 Speed Reduction. An automatic method of speed reduction for pump and fan motors, or air or water flow reduction, during less than full system load conditions, which assures energy savings through motor power input reduction, shall be employed for any air system that exceeds a total system static pressure of 2.0 inches water gauge, and any water system that exceeds a total water system pressure equivalent to a 50 foot head of water.*

**Section 503.5 Balancing. Add a new second sentence to read:** *For structures with a floor area greater than 5000 square feet with forced-air climate control, balancing shall be performed or included as part of a commissioning process from the design and construction phase. Certification and results of the balancing shall be submitted to the jurisdiction, the owner, and the designer of the project.*

**Section 503.8.1. Strike entire section. Section should read as follows:** *All ducts shall be leak tested in accordance with this standard. The tested rate of air leakage is not to exceed 3% of conditioned floor area in CFM at 25 pascals (0.1 inches WC). A representative of the developer and/or builder will perform a field inspection and leakage test of the ductwork before drywall installation. The field representative will certify successful completion of this test.*

**Section 503.9 Piping Insulation.** Delete exceptions 2 and 4, and renumber exception 3 to 2.

**Section 504.2.1.1. Electric Water Heaters. Revise to read:**

*All Automatic electric storage water heaters shall have a standby loss not exceeding 4.0 watts/ft$^2$ (43W/m2) of tank surface or 43 watts, whichever is greater, when tested in accordance with Standard RS-5 listed in Chapter 8 and calculated at a 30 Degrees F. temperature difference.*

**Section 504.5.1. Add the following at the beginning of the paragraph:**

*All recreational swimming pools and spas shall utilize solar energy as the only water heating source. Medical and rehabilitation pools smaller than 3,000 gallons water capacity shall use solar energy as the primary water heating source, with a new energy source permitted as backup.*

**Add a new Section 504.5.4:**

*Section 504.5.4 Recirculating Systems. Recirculating systems shall be provided with time clocks as required in Section 504.5.3, switches as required in Sec. 504.6, and pipe insulation as required in Section 504.7.*

**Section 504.8 Conservation of Hot Water. Delete in its entirety and add:**

*Plumbing fixtures shall meet the following maximum usage requirements unless special requirements dictate otherwise:*

- *Water closets: 1.6 gallons per flush.*
- *Kitchen showers and lavatory faucets: 3 gallons per minute.*
- *Urinals: 1 gallon per flush.*

**Add a new section 506 to read:**

*Section 506. Energy Consumption—Other Than Electrical. In multifamily dwellings , provisions shall be made to determine the energy consumed by each tenant by separately metering individual dwelling units or tenant spaces.*

# CHAPTER 6

## Residential Building Design by Acceptable Practice

**Section 602.2.1 Walls.** Add the following paragraph to read:

*602.2.1.1 Wall assemblies. Exterior walls shall be constructed to meet a minimum composite R-Value of 19 including air films. The following assemblies are deemed to meet this requirement:*

1. *Nominal 2 x 6 wood-frame construction insulated with R-19 batts or applied blown-in process custom cavity filling insulation.*

2. Nominal 2 x 4 wood-frame construction insulated with R-13 batts or applied blown-in process custom cavity filling insulation and rigid insulation board on the exterior with an R-Value of not less than 3.2, with a nominal 4 inch brick veneer.

3. Nominal 2 x 4 wood-frame construction insulated with R-13 batts or applied blown-in process custom cavity filling insulation and rigid insulation board on the exterior with an R-Value of not less than 6.0.

4. Earthen material or solid masonry, at least 4 inches in thickness with insulation with an R-Value of not less than 9.0 applied to the exterior.

5. Earthen material or solid masonry, at least 12 inches in thickness or 8 hours time lag.

6. Straw bales at least 14 inches in thickness.

7. Log or solid wood construction with an average thickness of 12 inches.

8. Log or solid wood construction with an average thickness of 6 inches with nominal 2 x 4 frame construction insulated with R-13 batts or loose fill insulation on the inside.

9. Structural insulated panels with a minimum R value of 25.

Add the following exceptions after the last paragraph:

*Exceptions:*

1. Doors whose area and U-value are considered as glazing in Section 602.2.1 shall be exempt.

2. One exterior swinging door with a maximum area of 32 square feet may be installed for ornamental, security or architectural purposes and be exempt from these requirements.

Add new subsections to read:

**602.2.1.1 Exterior door area.** For doors containing at least 20 percent of the door area in glazing, the glazing area shall be subtracted from the door area for the purpose of determining the U-value of the door.

**602.2.1.2 Exterior door U-value.** All exterior doors shall have a maximum area weighted U-value not exceeding that prescribed in Table 102.3b.

**602.2.1.3 Glazing.** All glazing facing 20–165 degrees or 195–340 degrees shall have a minimum summer shade or shading coefficient of 0.39. All glazing facing between 165–195 degrees shall have a minimum summer shade or shading coefficient of 0.5 or less. This may be accomplished by the use of overhangs, covered porches, tinted glazing or other approved methods. Whole-unit glazing maximum U-value shall be determined by the following equation or Table 602.2.1.3:

$$U = 5 / [(\text{Glazing Percentage}) - 5] \text{ or } R = [(\text{Glazing Percentage}) - 5] / 5$$

where Glazing Percentage = 100 x (glazing area / floor area)

*Table 602.2.1.3*

| Percent of floor area | Window R value |
|---|---|
| 10 | 1.0 |
| 12 | 1.4 |
| 14 | 1.8 |
| 16 | 2.2 |
| 18 | 2.6 |
| 20 | 3.0 |
| 22 | 3.4 |
| 24 | 3.8 |
| 26 | 4.2 |
| 28 | 4.6 |
| 30 | 5.0 |

***602.2.1.4 Summer Ventilation.*** Buildings utilizing thermal-mass (items 4 and 5, in Section 602.2.1.1) shall be provided with a means of venting to the outside at night during the months of May through October to avoid overheating. Operable windows totaling at least 20 percent of the total glazing area, located for effective cross-ventilation and ceiling fans or a whole-house fan sized to provide 10 air changes per hour may be used.

**Section 602.2.2 Roof/ceiling. Add a new subsection to read:**

***602.2.2.1 Roof/ceiling assemblies.*** Ceilings below ventilated attic spaces and single rafter vaulted ceilings shall be constructed to meet a minimum composite R-value of 38 including air films. The following assemblies are deemed to meet this requirement:

1. Wood-frame assembly insulated with R-38 batts or loose-fill insulation.

2. Earthen material or solid masonry at least 12 inches in thickness with glass fiber or foam insulation with an R-Value of not less than 22 applied to the interior.

3. Earthen material or solid masonry at least 24 inches in thickness with glass fiber or foam insulation with an R-Value of not less than 11 applied to the interior.

4. Earthen material or solid masonry at least 36 inches in thickness

5. Straw bales at least 14 inches in thickness.

6. Structural insulated panels with an R value of at least 38.

**Section 602.2.4. Slab-on-grade floors.** Delete 2nd paragraph and insert the following sentence: *At least 25 percent of the floor area in the living space shall be without carpet or an equivalent area of internally exposed mass.*

**Section 602.3. add the following sentence at the end:**

*An Air Leakage Warranty verifying a maximum of 0.35 ACH shall be provided to the homeowner per Section 502.3.*

**Section 603.2. HVAC equipment requirements, change to read:**

*... efficiency and power density requirements of section 503.2, 503.3 and 503.8.1.*

**Table No. 603.5. Minimum Pipe Insulation.** Revise by changing the numbers ~~1 1/2~~ to *1/2* in the third and fourth columns on the line "CHILLED WATER."

**Section 603.6. Add a new section to read:**

*603.6. Space Cooling. Buildings constructed under the provisions of this standard shall be permitted to use refrigerated air conditioning systems selected under the guidelines of the Air Conditioning Contractors of America (ACCA)Manual J Procedures, Specifically Sections 7-27, 7-28 and 7-29 at outside conditions of 105 degrees F. and inside conditions of 78 degrees F. Other provisions of this standard notwithstanding, air conditioning equipment shall have a minimum SEER of 12 or a minimum EER of 10. System sizing shall be determined by an analysis consistent with industry standards.*

**Section 603.7. Add a new subsection to read:**

*603.7. Mechanical Equipment and Water Heater Efficiency. The efficiency of water heating and mechanical equipment shall be in accordance with sections 101.4, 102.4 and 503 of this standard.*

Check List when using Chapter 6

| Section | Description | Check |
|---|---|---|
| 102.1.3 | "Insulation Installation Warranty" completed and signed. | |
| 102.4 | Minimum of 600 ft$^2$/ton. | |
| 104.1 | A method of utilizing solar energy (may overlap with 504.2). | |
| 503.2.4 | Recessed lighting fixtures when installed in building envelope are constructed to accept only lamps with efficacy greater than 40 lumens/watt. | |
| 503.8.1 | Duct Leakage Test passed before drywall and "Intermediate Verification and Warranty" form is signed. | |
| 101.4, 102.4 | Water heater system demonstrating compliance. | |

| | |
|---|---|
| 504.2.1.1 | Electric storage water heaters have a standby loss not to exceed 4 watts/ft$^2$ of tank surface or 43 watts, whichever is greater. |
| 504.5.1 | Pools and spas utilize solar energy as the only water heating source. |
| 504.5.4 | Recirculating system (if used) is installed on hot water line with a timer and pipe insulation. |
| 504.8 | Low water use plumbing fixtures conform with UPC. |
| 602.2.1.3 | Glazing facing between 195–340 & 20–165 degrees has a summer shade or shading coefficient of 0.39 or less. Glazing facing between 195–165 degrees has a summer shade or shading coefficient of 0.5 or less. |
| 602.2.1 | Wall assembly from table 602.2.1.1 |
| 602.2.1.2 | All exterior doors but one (maximum of 32 ft$^2$) to be R-5. |
| 602.2.1.3 | Minimum window U value as per Table 602.2.1.3. |
| 602.2.1.4 | Buildings utilizing thermal mass walls have minimum 20% operable windows or ceiling fans or a whole house fan (10 ACH). |
| 602.2.2.1 | Minimum composite R value of 38 for roof/ceiling. |
| 602.2.4 | Minimum of 25% of floor area without carpet or equivalent area of other internally exposed mass. |
| 602.3 | "Air Leakage Warranty" verifying maximum of 0.35 ACH. |

## SECTION 604

## SERVICE WATER HEATING

**Section 604.1.2.3 Swimming Pools *and Spas*.** Add the following at the beginning of the first paragraph:

*All recreational swimming pools and spas shall utilize solar energy as the only water heating source. Medical and rehabilitation pools smaller than 3,000 gallons water capacity shall use solar energy as the primary water heating source, with a new energy source permitted as backup. When permitted, spa and P̶pool ... remainder unchanged.*

**Section 604.2 Water heaters, storage tanks and boilers.** Add a sentence to read:

*Shall comply with Section 101.4 and 102.4 of this standard.*

**Section 604.4 Conservation of hot water.** Delete in its entirety and add: *Plumbing fixtures shall meet the following maximum usage requirements unless special requirements dictate otherwise:*

- *Water closets: 1.6 gallons per flush.*
- *Kitchen showers and lavatory faucets: 3 gallons per minute.*
- *Urinals: 1 gallon per flush.*

## CHAPTER 7

## Building Design for All Buildings Other Than Residential Buildings

**Section 702.1 Basic Requirements.** Revise to read as follows:

**702.1 Basic Requirements.** Building designs shall meet the requirements of *Section 101.4 and 102.4 of this standard* as well as the requirements of Sections 5.4, 6.4, 7.4, 8.4, 9.4, 10.4, 11.4 and 12.4 in Standard RS-22 listed in Chapter 8.

**Section 702.2 Additional requirements.** Add the following before the first exception:

*Buildings otherwise meeting the requirements of Chapter 4, 5 or 6 of this standard may employ the following performance requirement substitutions of Standard RS-22, listed in Chapter 8.*

Remainder unchanged.

**Add a new section to read:**

### SECTION 703

### ELECTRICAL REQUIREMENTS

*703.1 Lighting Power Budget. The maximum lighting power density (LPD) for any building may be calculated by either the Complete Building Method or the Area Category Method, according to the following table:*

*TABLE 703.1*

*Complete Building Method*

| Building Type | Max. Lighting Power Density (W/sf) |
|---|---|
| General commercial or industrial work buildings | 0.8 |
| Grocery stores | 1.2 |
| Industrial or commercial storage buildings | 0.5 |
| Medical buildings and clinics | 1.0 |
| Office buildings | 1.0 |

| | |
|---|---|
| Religious worship, auditorium/convention centers | 1.3 |
| Restaurants | 1.0 |
| Retail and wholesale stores | 1.3 |
| Schools | 1.2 |
| Theaters | 1.0 |
| All others | 0.5 |

*Area Category Method*

| Area Type | Max. Lighting Power Density (W/sf) |
|---|---|
| Auditorium | 1.3 |
| Bank/financial institutions | 1.2 |
| Classrooms | 1.3 |
| Convention/conference/meeting centers | 1.0 |
| Corridors, restrooms, support areas | 0.5 |
| Dining | 0.8 |
| Exhibit | 1.5 |
| General commercial and industrial work | 0.8 |
| Grocery | 1.3 |
| Hotel function | 1.5 |
| Industrial and commercial storage | 0.4 |
| Kitchen | 1.5 |
| Lobbies: Hotel lobby | 1.5 |
|   Main entry lobby | 1.0 |
| Malls, arcades, and atria | 0.8 |
| Medical and clinical care | 1.2 |
| Office | 1.0 |
| Precision commercial and/or industrial work | 1.3 |
| Religious worship | 1.4 |
| Retail sales, wholesale showrooms | 1.4 |
| Theaters: Motion picture | 0.7 |
|   Performance | 1.0 |

*For any building greater than 5,000 square feet in area, and for all non-residential buildings, the following shall apply:*

Lighting design shall comply with current IESNA guidelines and application notes, with selection of the median illuminance as the target optimum, not the minimum. In all areas designated higher than IESNA category C, specified illuminance shall be on task, with ambient illuminance generally task/3. LPD shall in no case exceed the values in Table 703.1. In all areas, lighting targets shall be achieved by the most energy efficient technology which meets the following requirements:

- CRI greater than 80 for work areas, greater than 75 for all areas.
- Color temperature less than 3600 K, except in retail display, excluding tasklights.

Technologies include but are not limited to:

- Incorporation of natural daylight, and daylight-supplementing artificial light
- Area lighting by high-CRI straight tube fluorescent with specular reflectorized interior luminaire. For applications which do not involve frequent switching, use fully electronic instant start 4-lamp parallel-circuit ballasts. Daylight-dimming via photocell sensor/controller.
- Task lighting by 13 watt lamps (or smaller), with laterally offset placement so as not to cause direct or veiling glare.
- Conference rooms to have continuous architectural-dimming (to 10%) fluorescent luminaries in place of incandescent can downlights, or else can lights controlled by center-off double throw switch which prevents simultaneous operation of area lights and downlights.
- Reflectorized technologies for interior luminaries.
- Areas with visible daylight may use continuous-dimming photocell-controlled variable light output devices except those controlled by non-defeatable occupancy sensors. Daylighting contribution to be considered in calculation of IESNA target illuminance.

Prohibited technologies include:

- Tungsten filament incandescent except quartz-halogen
- T12 fluorescent
- VHO and SHO lamps
- Mercury vapor lamps
- U-shaped lamps (except CFLs <20W)
- Small-cell paracube grates and large cell paracube fixtures where the lamp is not centered into the cell
- Magnetic fluorescent ballasts
- Series-circuit ballasts

Occupancy Sensors shall be specified and installed in accordance with EPA Green Lights guidelines, with control technology appropriate to the application. Areas with video display terminals shall be primarily illuminated by task lighting, use of overhead luminaries in each space to be evaluated for Visual Comfort Probability. Visual acuity factors are to be treated as minus weighting factors.

## EXHIBIT 2 CERTIFICATION

The Community of Civano, LLC has reviewed the plan for _____(address), plan number _____to determine compliance of that plan with the conservation requirements for Civano as set forth herein.

The Community of Civano, LLC certifies it is familiar with and the plans meet the requirements of Memorandum of Understanding between the City of Tucson and the Community of Civano, LLC dated June 9, 1998 as indicated below:

\_\_\_\_\_ Section 5.3.1.1 of the Memorandum of Understanding for Residential Buildings or the code requirements of the Sustainable Energy Standard attached to the Memorandum of Understanding.

\_\_\_\_\_ Section 5.3.1.2 of the Memorandum of Understanding for Commercial Buildings or the code requirements of the Sustainable Energy Standard attached to the Memorandum of Understanding.

Compliance with section 5.3.1.1 or 5.3.1.2 has been determined by:

\_\_\_ Prescriptive Method

\_\_\_ Component Method

\_\_\_ Systems Method;

Date:

Signature:

On behalf of the

Community of Civano, LLC

# B

# REVISED SUSTAINABLE ENERGY STANDARD

### SUSTAINABLE ENERGY STANDARD

**For The International Energy Conservation Code, 2003 Edition**

**Regionally specific for the Tucson Metropolitan Area**

The following modifications to the International Energy Conservation Code, 2003 Edition are deemed to be a sustainable energy standard:

*(Editorial Note: This Energy Standard has been updated from the original Sustainable Energy Standard, CABO Model Energy Code, 1995 Edition, dated 4-22-98. While this standard may be beneficial for many regions it is specific to the Tucson Metropolitan area. This standard is not to conflict with the previous Sustainable Energy Standard or amendment to locally adopted codes.)*

*Chapter 1—Administration and Enforcement*

**Section 101.1. Title.** Delete brackets and "Name of Jurisdiction."

**Section 101.2. Scope.** Add a paragraph to read:

The calculated target annual energy consumption of the building lighting, mechanical system, and domestic hot water heating shall be less than 50% of the energy required by the ANSI/ASHRAE/IESNA Standard 90.1-2001 without amendments for the purpose of calculating the minimum base case, otherwise buildings must also meet the adopted International Energy Code of this jurisdiction. In addition, the minimum displacement goal of energy by solar devices is prescribed as a function of residential bedrooms at 550kWh/br/yr. Displacement for other structures is prescribed in tables relating displacement goals as a function of the buildings' use and occupancy. Buildings that show proof of LEED registration at the silver level with a LEED Accredited Professional as part of the design team shall be deemed compliant with this standard. New buildings must achieve a minimum of 7 points from LEED credits EA1 (Optimize Energy Performance) and EA2 (Renewable Energy). Existing buildings must achieve 9 points from LEED credits EA1 and EA2. Commercial buildings shall demonstrate that 5% of the total annual building lighting, mechanical system and domestic hot water heating energy consumption is offset by the use of solar energy for all methods of compliance.

**Exception:** For each 5% of building lighting, mechanical system and domestic hot water heating energy budget that is offset with co-generation the solar requirement may be decreased by 1% to a minimum of 1% solar energy utilization.

**Section 102.4.** Delete after "of" and add: , and there shall be a verification of proper installation of insulation before drywall installation, and the completion of the "Insulation Installation Warranty" and signature by a representative of the developer and/or builder.

**Section 102.6. Equipment.** Add a new subsection to read:

**102.6.1.** Residential buildings constructed under the provisions of this standard shall be permitted to use refrigerated air conditioning systems selected under the guidelines of the Air Conditioning Contractors of America (ACCA) Manual J Procedures, Specifically Sections 7-27, 7-28 and 7-29 at outside conditions of 105 degrees F and inside conditions of 75 degrees F. Other provisions of this standard notwithstanding, air-conditioning equipment shall have a minimum SEER of 12.5.

**102.6.2. Evaporative cooling.** Evaporative cooling may be used for cooling or to reduce air conditioning requirements but may not be used as the method of compliance to this standard except for commercial buildings that use evaporative cooling as an economizer cycle on a refrigeration or air-conditioning application. Duct leakage through the evaporative device shall be minimized during air-conditioning and heating modes of operation.

**102.6.3. Water Heating.** The following service water heating systems are the only methods acceptable:

a. Solar water heaters.

b. Instant gas or electric water heaters.

c. Heat pump electric water heaters.

d. Heat recovery water heaters from air conditioning or other sources.

e. Gas water heaters exceeding 80% efficiency.

f. Passive solar with in-collector storage (ICS), thermal siphon and alike shall be installed with no more than a total of 20 linear feet of piping between the solar system and the storage tank.

**Exception:** Other methods acceptable to the authority having jurisdiction showing 50% reduction of water heating energy consumption.

Water heating systems that serve only hand sinks and/or a single mop sink may use a water heater with up to 20 gallons of storage.

**Section 104.1 General.** Add a sentence at the end of the paragraph to read:

Plans and specifications shall show the method of utilizing "beneficial use of solar energy."

Add a new section to read:

**Section 108–Wood-Burning or Gas Fireplaces and Wood Stoves**

**108.1. Wood-burning stoves.** Wood-burning stoves shall be labeled to show compliance with the U.S. Environmental Protection Agency (EPA) Phase II standards for particulate emissions during operation.

Catalytic stoves shall have an accessible, modular, replaceable catalyst element.

**108.2. Fireplaces.** Wood-burning fireplaces and gas fireplaces shall produce useful heat and be provided with a means of supplying 100% of the combustion air for operation from the outside, and shall limit particulate emissions to less than 7.5 grams per hour. All fireplaces shall be provided with a tight fitting glass door and a positive means of circulating the heated air in the occupied space.

**108.3. Solar Backup.** A wood-burning stove or fireplace shall be considered as providing the required space heating energy only when installed as backup energy for a solar-thermal collection system.

*Chapter 2—Definitions*

**Positive Cooling Supply.** Amend as follows: insert including evaporative cooling systems, between "cooling" and "deliberately."

Add the following new definitions:

**Civano.** A Tucson Solar Village, a model sustainable community; a vision of the future where resource consumption is reduced through more efficient technologies, use of solar energy and lifestyles which promote greater harmony and balance with the natural environment; a community in the spirit of the "Civano" period, a golden era of the Hohokam culture which balanced natural resources and human needs; incorporates and demonstrates strategies for achieving more sustainable development.

**Sustainable Development.** "Development that meets the needs of the present without compromising the ability of future generations to meet their needs." (UN World Commission on the Environment and Development)

**Beneficial Use of Solar Energy.** The following devices/methods may be used to demonstrate compliance:

- Solar space heating systems.

- Trombe wall or clear view collectors for space heating.

- Solar photovoltaic systems.

- Solar thermal/electric power generating systems, including stand-alone and grid connected parabolic trough and dish Stirling.

- Solar day lighting system using glazing and controls to turn off electric lights or dimming of lighting.

- Solar day lighting systems specifically designed to capture and redirect visible solar energy while controlling infrared energy (conventional skylights are specifically excluded) for at least one half of the non-bedroom space.
- Passive building heating for the winter through the use of optimum window shade structures and orientation.
- Solar water systems for domestic water heating or space heating.
- Solar pool or spa water heating–see also 504.5.

**Power Density.** The total connected power load of all components of a building system, including all auxiliary components and circuitry, without regard to the timing, scheduling, or control of their operation, in w/ft$^2$ or Btu-h/ft$^2$.

**Site Energy.** Energy, other than recovered energy, utilized for any purpose on the site.

Source energy consumption shall be determined by multiplying the site energy usage in kBtu-h per square foot by the following factors:

| Site Energy | Factor |
| --- | --- |
| Electric | 3.10 |
| Gas | 1.11 |
| Wood | 1.00 |
| Solar (amount of displaced electric or gas) | 0.00 |

**Bedroom.** A room including clothes closets that may be used for sleeping purposes.

## Chapter 3—Design Conditions

Table 302.1 Exterior design conditions: Revise the table as follows:

**Table 302.1**
**EXTERIOR DESIGN CONDITIONS**

| WINTER | DESIGN DRY BULB TEMP. | 32°F |
| --- | --- | --- |
| SUMMER | DESIGN DRY BULB TEMP. | 104°F |
|  | DESIGN WET BULB TEMP. | 66°F |
| DEGREE DAYS HEATING |  | 7000 |
| DEGREE DAYS COOLING |  | 2814 |
| CLIMATIC ZONE |  | 14A |

## Chapter 4—Residential Building Design by Systems Analysis and Design of Buildings Utilizing Renewable Energy Sources

**Section 402.1 Analysis Procedure.**

Add a sentence at the end of the first paragraph to read: Domestic hot water energy use shall be calculated separately from glazing systems, heat storage, thermal envelope and space conditioning equipment and shall meet the energy reduction percentages of section 101.2.

**Section 402.5. Calculation Procedure** Add:

**402.5.4** The occupied mode shall be not less than 10 hours in a 24-hour period.

**Section 402.6. Documentation.** Add to the last sentence to read: Chapter 4 and that the derived proposed design is a minimum of 50 percent of the standard design.

## Chapter 5—Residential Building Design By Component Performance Approach

**Section 502.1.2.** Revise Title to "Masonry or Earthen Materials" and the note by placing a comma after "masonry," and adding "earthen materials," between "masonry" and "or."

**Section 502.1.5.** Change solar heat gain coefficient from 0.4 to 0.35.

Note: solar heat gain coefficient = shading coefficient x 0.87.

**Table 502.2.** Add the following values:

### Table 502.2
### HEATING AND COOLING CRITERIA

| ELEMENT | MODE | TYPE A-1 BUILDINGS $U_o$ | TYPE A-2 BUILDINGS $U_o$ |
|---|---|---|---|
| Walls | Heating or cooling | 0.11 | 0.17 |
| Roof/ceiling | Heating or cooling | 0.026 | 0.026 |
| Floors over unheated spaces | Heating or cooling | 0.05 | 0.05 |
| Heated slab on grade | Heating | R-Value 8 | R-Value 8 |
| Unheated slab on grade | Heating | R-Value 0 | R-Value 0 |
| Basement wall | Heating or cooling | U-Factor 0.095 | U-Factor 0.095 |
| Crawl space wall | Heating or cooling | U-Factor 0.06 | U-Factor 0.06 |

Delete all footnotes.

**Section 502.2.1.4 and 502.2.3.4. Slab-on-grade floors.** Delete in its entirety.

**Table 503.2** Change the values in the fourth column (Minimum Performance) from HSPF ~~6.8~~ and ~~6.6~~ to 7.0 and 7.0 respectively, change SEER values in the fourth column (Minimum Performance) from ~~10~~ and ~~9.7~~ to 12.5 and 12.5, respectively

**Section 502.3.3.** Recessed lighting fixtures, amend as follows:

**502.3.3. Recessed lighting fixtures.** When installed in the building envelope, recessed light fixtures shall be sealed to prevent air leakage into or from the conditioned space.

**Section 503.3.3.1. Piping Insulation.** Delete exceptions 2, and 3.

**Section 503.3.3.4.2.** Add to end: All low pressure ducts shall be leak tested in accordance with this standard. The tested rate of air leakage is not to exceed 3% of conditioned floor area in CFM at 25 pascals (0.1 inches WC) prior to drywall and air handling equipment installation. A representative of the developer and/or builder will perform a field inspection and leakage test of the ductwork before drywall installation. The field representative will certify successful completion of this test.

**Section 504.3.** Add the following at the beginning of the paragraph:

All recreational swimming pools and spas shall utilize solar energy as the only water heating source. Medical and rehabilitation pools smaller than 3,000 gallons water capacity shall use solar energy as the primary water heating source, with a new energy source permitted as backup.

**Section 504.4.** Delete "conveniently," place a period after "automatically" and delete the rest of the sentence.

**Section 505.3. Lighting fixture efficacy.** Amend as follows:

**505.3. Lighting fixture efficacy.** All general purpose lighting fixtures in kitchen, laundry room, utility room, equipment room, and garage, and those that are required by other codes at entries on the exterior of buildings shall be so constructed as to accept only lamps with efficacy greater than 40 lumens/watt.

**Exception:** Those fixtures designed for spot or flood type lamps and those fixtures controlled by a permanently installed dimmer.

**Section 505.4. Exterior lighting fixture controls,** amend as follows:

**505.4. Exterior lighting fixture controls.** Exterior lighting fixtures shall be controlled by a time switch with astronomic adjustment or a photo sensor. A standard time switch may be incorporated with the photo sensor to turn the lights off at a desired time before dawn. All time switches shall incorporate a minimum 2 hour carry through of the program.

**Add a new section 506 to read:**

**Section 506. Energy Consumption–Other Than Electrical.** In multifamily dwellings, provisions shall be made to determine the energy consumed by each tenant by separately metering individual dwelling units or tenant spaces.

### Chapter 6—Simplified Prescriptive Requirements

For this procedure use climate zone 15 (7,000 HDD)

602.2 (same as 502.1.5.)

### Chapter 7—Building Design For All Commercial Buildings

Adopt ASHRAE 90.1 2004 without amendments.

### Chapter 8—Design By Acceptable Practice For Commerical Buildings

Delete entire chapter.

### Chapter 9—Climate Maps

Deleted for clarity.

# BIBLIOGRAPHY

Al Nichols Engineering, Inc. *Energy and Water Use in Tucson: January 2007–December 2007.* 13th annual IMPACT Process Civano Energy and Water Audit Report, Tucson: Al Nichols Engineering, Inc., 2008.

Allenby, Braden. *Reconstructing Earth: technology and environment in the age of humans.* Washington, DC: Island Press, 2005.

Armstrong, Susan J., and Richard G. Botzler. *Environmental Ethics: Divergence & convergence.* New York: The McGraw-Hill Companies, Inc., 2004.

Ashworth, John H., and Jean W. Neuendorffer. *Matching Renewable Energy Systems to Village-Level Energy Needs.* Prepared for the US DOE, Golden: Solar Energy Research Institute, 1980.

Butler, David. "Creating Energy Efficient and Healthy Homes." *RESNET.* David Butler, 2009. 13.

Butti, Ken, and John Perlin. *A Golden Thread: 2500 years of solar architecture and technology.* New York: Van Nostrand Reinhold Limited, 1980.

Civano Institute; City of Tucson; Pima County; Arizona Energy Office; Metropolitan Energy Commission. *Community Workshop Program Report for the Solar Village.* Progress/Educational, Tucson: Civano Institute, 1990.

Community of Civano, LLC. "Civano: Design principals and guidelines." *Civano.* Tucson: Community of Civano, LLC, February 1998.

Congress for the New Urbanism. *Charter of the New Urbanism.* New York: The McGraw-Hill Companies, Inc., 2000.

Congress for New Urbanism, Natural Resources Defense Council, U.S. Green Building Council. *LEED for Neighborhood Development Rating System: Pilot Version.* U.S. Green Building Council. 2007. Accessed from www.usgbc.org 06/2007.

Fatovich, Sandra, and Lupita Rios. "An Overview of The Open Public Meeting Law." *The Open Public Meeting Law.* Tucson: Office of the City Clerk, Tucson, Arizona, June 27, 1997.

Field, Barry C. *Environmental Economics: Second edition.* New York: The McGraw-Hill Companies, Inc., 1997.

Geller, Howard. *Policies for a More Sustainable Energy Future.* Policy Reform, Washington DC: American Council for an Energy-Efficient Future, 1999, 58.

Hawken, Paul, Amory Lovins, and L. Hunter Lovins. *Natural Capitolism.* Newport: Back Bay Books, 2000.

Kitchin, Rob: Tate, Nicholas, J. *Conducting Research into human Geography.* London: Pearson Education Limited, 2000.

Ludwig, Art. *Branched Drain Grewywater Systems.* Santa Barbara: Oasis Design, 2000.

—. *Builder's Greywater Guide: Installation of greywater systems in new construction and remodeling.* Santa Barbara: Oasis Design, 1995.

—. *Create and Oasis with Greywater: Choosing, building and using greywater systems.* Santa Barbara: Oasis Design, 1994.

McNeill, J. R. *Something New Under the Sun: An environmental history of the twentieth-century world.* New York: W.W. Norton & Company, Inc., 2000.

Merchant, Carolyn. *Radical Ecology: The search for a livable world.* New York: Routledge, Chapman & Hall, Inc., 1992.

Moughtin, Cliff. *Urban Design: Green dimensions.* Jordan hill: Architectural Press, 1996.

NAHB Research Center. *Tucson Solar Village Emerging Housing Technologies.* Progress Report, Upper Marlboro: National Association of Home Builders, 1991.

Nordhaus, Ted, and Michael Shellenberger. *Breakthrough: From the death of envrionmentalism to the politics of possibility.* New York: Houghton Mifflin Company, 2007.

Old Pueblo Archaeology Center. "It was the Hohokam who Coined the Name "Civano"." *The Town Crier*, September/October 2007: Cover, 5.

Petersen, Jacuzzi & Green, Inc. *Civano: A study to determine the keys of success for developing the community of Civano.* Final Market Study Report to Community of Civano, LLC, Phoenix: PJ&G, Inc., 1997.

Sonoran Institute. *Building from the Best of Tucson.* Tucson: Sonoran Institute, 2001.

Spring, Cari Ph.D. *When the Light Goes On.* Tucson: Emerald Resource Solutions, 2001.

Subcommittee to the Tucson/Pima Energy Code Committee. "Sustainable Energy Standard." *Mayor and Council Resolution No. 18082.* Tucson: Tucson Mayor and Council, July 06, 1998.

Tucson Institute for Sustainable Communities. *Sustainable Design: A planbook for sonoran desert dwellings.* Tucson: Tucson Institute for Sustainable Communities, 1999.

Vivian, John. *Practical Homesteading.* Emmaus: John Vivian, 1975.

Witmer, Gallagher. *Impact System Monitoring Report #13.* IMPACT Process Civano Monitoring Report, Tucson: Gallagher Witmer, Architect LLC, 2008.

Wymore, Wayne A. PHD, and Melanie Wymore. "Civility: The Civano village environment system for the city of Tucson." *Civility.* Tucson: SANDS: Systems Analysis and Design Systems, February 24, 1993.

Yudelson, Jerry. *Choosing Green: The homebuyer's guide to good green homes.* Gabriola Island: New Society Publishers, 2008.

—. *Green Building Through Integrated Design.* New York: The McGraw-Hill Companies, Inc., 2009.

# INDEX

## A

Arizona Solar Energy Commission (ASEC) Land Committee, 1, 8–9

## B

Barna, Richard (interview), 203–209
barrier to sustainability, 45, 156
base loads
　calculation, 83–84
　plug or process, 83–84, 86
bedroom communities, 11, 63–64, 69, 190
biome, 14, 235
blower-door tests, 212
brownfield (high-priority), 16–17
Buntin, Simmons (interview), 180–183
Burton-Hampton, Gina (interview), 201

## C

candidate species, 14
centers
　commercial, 62–63, 67, 185, 188–192, 229, 231
　neighborhood, 64–65, 93, 102, 132, 151, 167, 179–180, 189, 196, 241
charrette
　design, 54, 183, 232
　process, 74, 76, 101–102, 155, 191
Civano Memorandum of Understanding, 21–25, 81, 92–93, 131, 238, 251–286
cogeneration, 8
commons (tragedy of), 14, 38–39, 46
Cook, Bob (interview), 42, 71–72, 83, 105, 109–110, 173
cooling degree days, 36, 108
Crockett, Doug (interview), 114–120

## E

Elwood, David (interview), 41
energy
　solar, 3, 4, 23, 43, 45, 60, 71, 82, 86, 109, 145, 158–162, 188, 195, 212, 214, 222, 237
　source, 82, 98, 214

## G

geographic information systems (GIS), 59
greenwashing, 37, 147

## H

heating degree days, 36, 80, 108, 210–211, 214–215
HERS rating, 194, 207–208
Houghton Area Master Plan (HAMP), xiv, 12–14, 231–234
human scale, 17, 20, 61, 65–68, 127
hydrology, 5, 15, 24, 153, 183, 216

## K

Kelly, Kevin (interview), 131–135, 139–142
Kentlands, 9, 70

## L

land selection, 109
Laros, Jason (interview), 28, 35–36, 153, 166–167, 180, 184, 191–193
Laswick, John (interview), 121–129, 142
Leadership in Energy and Environmental Design (LEED), 2, 3, 12, 15–17, 61–63, 66, 95–97, 99, 114, 118, 124, 149, 167, 192, 194, 216–220, 222, 224, 226–227, 232, 238
lessons learned, 3, 33, 45, 49, 58, 62, 92, 102, 104–110, 120, 127, 141, 142

## M

manual J calculations, 212
Martinez, Hector (interview), 47–48
master plan, 51, 53–61, 64–65, 72–76, 102, 108, 132–135, 153, 180–181, 197–198, 232–236
Miller, John Wesley (interview), 3, 4, 40
mixed density, 8
mixed use, 63–65, 67, 69, 93, 112, 132, 141, 175, 217, 224, 231, 241
Moody, Wayne (interview), 25, 41, 53–54, 56–59, 64, 72–73, 229–230

## N

New Urbanism (Congress for the, Charter of the), 2, 11, 66, 192, 231

Nichols, Al (interview), 28, 35–37, 78–79, 102, 109, 112, 181–184, 187–188, 195, 197–198, 201, 210–212, 229, 233–234, 238

## P

permaculture, 15–16, 141
planning
  general, 1, 4, 5, 7–10, 13–14, 24–25, 27, 31, 37, 40–42, 45–48, 102, 108, 148, 151, 175, 199, 201, 209, 230
  permit, 4, 8–9, 47

## R

Rayburn, Lee (interview), 74–75, 102–104, 108–113, 120, 175, 177, 188–191, 193–198
rezoning, 5, 13, 31, 47, 143, 149
riparian, 14–15
Rollins, Paul (interview), 53–54, 70–72, 229

## S

Scott, Shirley (interview), 19, 32–35, 168, 236
Seaside, 8, 70
Shipley, Les (interview), 241
Singleton, Jim (interview), 142–143, 173, 210
site
  analysis, 4–7, 11, 13, 15–24, 150
  selection, 32, 48, 56, 59, 62–66, 68, 72–73, 90, 109, 111
social value composite analysis, 58
Sonoran Desert, 11, 14, 75, 150, 217, 230, 237
Sustainable Energy Standard (SES), 29, 32–33, 36, 48, 71, 83–84, 89, 114, 143, 157–158, 176, 185–186, 195, 207, 210–211, 214–215, 220, 222, 232, 238, 287–293

## T

Tucson/Pima County Metropolitan Energy Commission (MEC), 28, 40–42, 48, 54, 58, 71, 79, 114–115, 125–127, 184

## U

United States Green Building Council (USGBC), xiv, 1–3, 12, 96, 216–217, 232, 235
University of Arizona's Environmental Research Laboratory, 4, 43
urban sprawl, 12–13, 64, 68, 121, 134, 196

## V

variances, 31
vertically integrated company, 91, 186–187
Village Homes, 9, 109, 123, 146

## W

Whitmer, Gal (interview), 152–153, 166–167, 185, 192–193

## X

xeriscape, 90, 153, 258

## Y

Yoklic, Martin (interview), 146, 153–156

## Z

zoning maps, 31, 74